Praise for

THE GENIUS FACTORY

"[An] ebullient new contribution to the realm of weird-science non-fiction . . . perfectly pitched—blithe, smart, skeptical, yet entranced by its subject."
—*The New York Times*

"A riveting account of a truly bizarre episode in American history—Robert Graham's crusade to save the human race. David Plotz has written a superb book about the quest for genius, and, ultimately, family."
—MALCOLM GLADWELL,
author of *The Tipping Point* and *Blink*

"[Plotz] pulls off the tricky feat of taking readers on a trip both serious and silly. . . . In between the alarming and the absurd, we also get something more, something unexpected: an ongoing, fascinating and deeply felt meditation on fatherhood and family."
—*Salon*

"The human story is painful and brilliantly related. . . . This is not just another local tale of American freakery, this is the story of a fundamental change in our attitudes to reproduction. Unpretentious, well organised, simply and readably told, this is a fine book about the human spirit and its indomitable pursuit of error."
—*The Sunday Times* (London)

"A wonderfully readable and eye-opening account . . . By giving readers the case study of a serious—and failed—effort to engineer a better human race, Mr. Plotz brings the discussion back down to earth, where it belongs. Tense, hilarious, and touching."
—*The Wall Street Journal*

THE GENIUS FACTORY

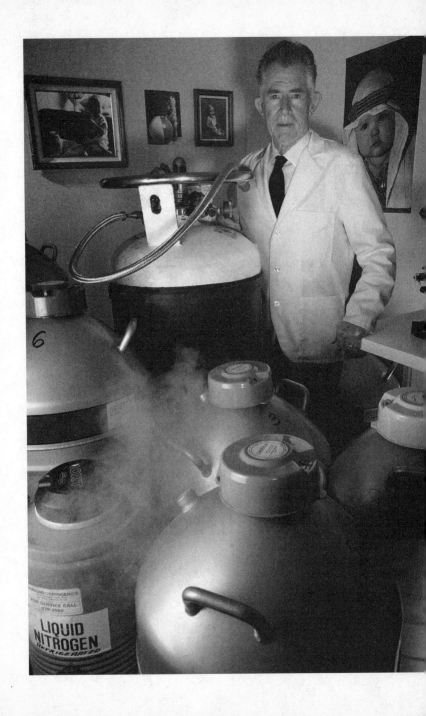

THE
GENIUS
FACTORY

THE CURIOUS HISTORY OF THE
NOBEL PRIZE SPERM BANK

DAVID PLOTZ

RANDOM HOUSE TRADE PAPERBACKS

NEW YORK

FOR HANNA

2006 Random House Trade Paperback Edition

Copyright © 2005 by David Plotz

Published in the United States by Random House
Trade Paperbacks, an imprint of The Random
House Publishing Group, a division of
Random House, Inc., New York.

RANDOM HOUSE TRADE PAPERBACKS and colophon
are trademarks of Random House, Inc.

Originally published in hardcover in the
United States by Random House, an imprint of
The Random House Publishing Group, a division
of Random House, Inc., in 2005.

This book was inspired by a series of articles written by
the author for *Slate* magazine. While some of the same
stories are included in this work, they have been
greatly expanded from the original articles.

LIBRARY OF CONGRESS CATALOGING-IN-PUBLICATION DATA
Plotz, David.
The genius factory: the curious history of the Nobel
Prize sperm bank / David Plotz.
p. cm.
ISBN 0-8129-7052-7
1. Graham, Robert Klark. 2. Sperm banks—United
States. 3. Artificial insemination, human—United States.
4. Nobel Prizes. 5. Intellect—genetic aspects.
I. Title.
HQ761.P56 2005   362.17'83—dc22
2004051497

Printed in the United States of America

www.atrandom.com

1 2 3 4 5 6 7 8 9

CONTENTS

AUTHOR'S NOTE

Many of the names in this book are pseudonyms. I have changed the names of all the sperm donors except William Shockley, who publicly (and proudly) admitted his contribution to the Nobel bank. I have changed the names of all the Nobel sperm bank children (as well as those of their parents) except two, Victoria Kowalski and Doron Blake. Their names have been public for more than twenty years. Victoria Kowalski was the first Nobel sperm bank baby, and her parents sold the story of her birth to the *National Enquirer.* Doron Blake was the second child. He and his mother, Afton, have been giving interviews since he was two weeks old. I have also changed identifying details about some donors and children, notably their hometowns and professions.

That moment, the moment he learned the truth, Tom Legare thought: *This is when my old life stops and I have to start over. I am a new me. I have a new future.*

It happened in February 2001. Tom was fifteen years old. He was spending a long weekend in Springfield, Illinois, with his mom and his younger sister. His dad, Alvin, was on the road. He was a salesman, always traveling. Even if Alvin hadn't been working, he wouldn't have come with them. He didn't go on family vacations. That was the kind of father Tom felt he had always had: indifferent, gloomy, detached. He still "lived at home"—if that's what you call sleeping in the house a few days a month—but he had left his wife and kids. Or maybe they had left him.

Tom and his mother, Mary Legare, were eating breakfast in the hotel restaurant, at a table by the window. They were waiting for his sister, Jessica, who was upstairs sleeping. Tom and Mary were spoiling for a fight. What teenager and his mom aren't?

Mary said, "Thomas, I need to talk to you."

Only she called him Thomas.

"Thomas, you're at the point where you have to make decisions about what you're going to do with your life." She paused. "And, Thomas, I don't like the direction you're heading in."

He said nothing and scowled at her.

"I'm proud you're going to graduate a year early from high school. But I heard you talking about going to pro wrestling school, whatever that is. Thomas, that is *not* going to happen."

Tom sighed. It was 8 A.M. He was on vacation. He wanted to finish his Pop-Tarts. He wanted to play video games in the hotel game room. Besides, the wrestling school plan was a joke, something to talk smack about with his friends when they were practicing moves in his backyard ring. But Tom didn't say that. He didn't want to give his mom the slightest satisfaction. He knew that his silence would just annoy her more.

She kept going. "You can't go to wrestling school. You can't waste your life on that.

"Thomas, you can do better than that. I *know* you can do better."

When he heard her say that, Tom wondered if his mom's secret was going to come out. This wasn't the first time she had dropped that same hint, with the same emphatic certainty in that word "know": "I *know* you can do better." From the way she said it, he knew, just knew, she was keeping something from him.

Irritated by her hinting, Tom broke his silence.

"Why do you *know* I can do better, Mom? Why? If there is something you want to tell me, just tell me."

She thought for a moment, then asked, "Thomas, have you ever heard of the Repository for Germinal Choice?"

"No," he said.

"You know how I always told you that you didn't have to be like your dad? I don't want you to look at your father and say, 'This is what I have to look forward to.' You *need* to know that you have better potential than your dad, because you don't have your dad's genes."

Suddenly Tom wanted her to stop. Somehow he knew that if she kept talking his life would never be the same, that his childhood would be over. But he couldn't think of anything to say, and she continued.

"When we were first married, we wanted a baby and I couldn't get pregnant. We went to a doctor and he said that because of a Vietnam injury your dad couldn't have kids. Then the doctor told us about an amazing sperm bank in California—the Repository for Germinal Choice. All the donors were Nobel Prize winners. I thought that was really special."

"And that's where you come from, Thomas. Your dad was a Nobel Prize winner."

Tom didn't want his mom to think she had won the wrestling school argument, so he snapped, "This doesn't change a damn thing." But he knew he was lying. He knew it changed everything. Tom turned it over in his head: a sperm bank. A *Nobel Prize* sperm bank. Tom felt surprised and not at all surprised. It explained everything, he thought: *That's why me and Dad don't look alike. That's why we don't get along. That's why Dad treats me that way, keeps me at a distance.*

"Do you know who my—what do I call him?—my 'real dad' is?" Tom asked.

"They didn't tell us his name, just a code name. I know he was a brilliant scientist."

"Is Jessica my sister?"

"Yes, but she comes from a different donor, so she's your half sister. She doesn't know anything about this yet. She's not ready to know, so please don't tell her. Promise you won't tell her."

One last question: "Can I find my real dad?"

"No, you can't. I have a sheet of paper somewhere in the house about him. That's all I have. The sperm bank is closed, so you won't be able to find his name or what he does or who he is. I'm sorry."

Tom left the restaurant and wandered off to check his head. The next three days vanished for him. Even today he can't remember what he did or what he thought. When they returned home from vacation, to their Kansas City suburb, Tom stopped talking to his mother. He shared his new secret with his three best friends; he trusted them because they *weren't* family.

After a few more days, he started to feel better. It was pretty funny, he thought: One day you're a long-haired, fifteen-year-old slacker pulling Bs at a mediocre high school, dreaming of rapping and wrestling. The next day, you're Superkid. Every boy fantasizes at some point that his parents aren't his real parents, that his real dad is a king or a billionaire or a movie star. "Hey," Tom said to himself, "I really do have a secret dad, and he really is a Nobel Prize winner!"

Tom was an optimist by nature, and he told himself that only good things could come of it. His dad had wasted his life, and his mom had had to work her butt off for all that she had accomplished. But maybe it will be different for me, he thought. Now I have potential. I have genius genes.

Tom started drinking less, and stopped smoking pot. His grades ticked up. He did his homework, for a change. He was happier.

He felt oddly relieved. Over and over, he would say to himself: "I'm not related to Dad. I'm not related to Dad, thank God!" But he didn't tell his dad he knew the secret. Why bother? It wouldn't change anything. It certainly wouldn't make them closer.

It was the *other* dad that Tom thought about. Tom puzzled over what to call him, even what to call him in his own head. At first, Tom tried "my real dad," but that felt wrong. "My donor"—too clinical. "Biological father"—too cumbersome. "My other dad"—that sounded as if there were some complicated family dispute.

Tom played genetic "Who am I?" games with himself. What is the great gift *he* gave me? If *he* was such a great scientist, why am I so bad at science? I can't draw at all: Could he draw? I'm pretty good at math and history and English, and especially at music. Did all that come from *him*?

But when he realized he couldn't answer his own questions, Tom's curiosity soured to frustration. I don't know half of who I am, Tom thought, and I will never know unless I find him. Tom decided he could find his dad, no matter what his mom said. After all, how many Nobel Prize winners could there be? And how many of them looked like Tom?

Tom ransacked the house and finally discovered a thin manila folder buried in the back of his mom's file cabinet. It only deepened the mystery. Inside were ten sheets of white paper stapled together. The cover page read "Repository for Germinal Choice, Catalog of Donors." Each page had a code name at the top—a combination of a color and a number—and a brief description of the donor below. "Donor Turquoise #38" was brown-haired and blue-eyed—like me, Tom thought—"a top science professor at a major university" who was also "a professional musician." Was that his new dad? Or was it Donor White #6, "a scientist involved in sophisticated research" who enjoyed "reading history"? Or Donor Coral #36, who had an IQ of 160 and "excels in mathematics"? Or Donor Yellow/Brown #22, "one of our very great scientists," who was fond of mountaineering?

Tom was baffled, but his mom gave no help. The catalog didn't jog Mary's memory. She couldn't recall what donor she had chosen. She shrugged and advised Tom to restrain his curiosity and stop looking. Tom seethed at her: This was the biggest decision of *my* life, and you can't remember anything—not a code name, not a profession, nothing?

After festering for a while, Tom cleared his head and thought, *Okay, I can do this.* He studied the folder again. It also contained some of the correspondence between his mom and the Repository for Germinal Choice. The stationery letterhead listed the sperm bank's trustees and directors. One was Robert Graham. Tom knew from an Internet search that Graham had owned the sperm bank, so he couldn't be Tom's father. Two of the other officers were women, so they were out. Then he saw Jonas Salk's name. Salk was listed as a trustee of the Nobel Prize sperm bank. Jonas Salk, the inventor of the polio vaccine—Tom had read about him in school. Tom leafed through the donor catalog again and reread the description of Donor Yellow/Brown #22. Yes, Jonas Salk was certainly "one of our very great scientists." Jonas Salk. Tom let it settle in. Maybe Jonas Salk is my father.

Tom and I were looking for each other, but we didn't know it yet. As he pored over the donor catalog, I was several months deep into my own Nobel Prize sperm bank obsession.

My fixation started with my dad. People always say their relationship with their parents is "complicated." My relationship with my father has never been complicated. It's purely adoring. My father is a doctor, a rheumatologist who has spent more than thirty years— my whole life—directing the same research lab at the National Institutes of Health. He studies and treats patients with some of the cruelest illnesses there are—baroque muscle diseases that gnarl limbs, steal strength, and cripple kids.

Most people find my father forbidding when they first meet him. He is tall and thin, with a wide, round face and a grand nose. His voice is a commanding rumble. If he says, "David, this must be the best bread in New England"—and he is prone to such preposterous overstatements—he makes it sound as though you'd be a fool to contradict him. But anyone who talks to him for five minutes recognizes that he is sweet, gentle, goofy, humane, and wise. My friends go to him for advice. I go to him for advice. He made growing up easy for me. How could I become a good man? I watched him and followed.

My father is not a religious man. His creed is scientific. He believes in the primacy of rational thought over superstition. He considers science a kind of priesthood, by which he means that scientists have a sacred obligation to seek truth rather than profit or glory, and by which he also means that scientists should never, ever, ever make fools of themselves.

Which is how the Nobel sperm bank entered my life. It was early 1980; I had just turned ten years old. My father and I were sitting in the breakfast room, reading *The Washington Post* over cinnamon toast and orange juice. (I had the comics section.) All of a sudden, my father poked at his newspaper and erupted. "That's just the silliest idea I have ever heard of! What's wrong with Shockley?"

"Shockley!" The name jarred my brain.

I asked my father what was wrong. He tried to explain to me that William Shockley was a great physicist who had invented the transistor and won the Nobel Prize and was now involved in some Nobel sperm bank, and that it was a moronic idea, the sort of thing Hitler would have tried. I knew what the Nobel Prize was. But I didn't know what the transistor was. I didn't know who William Shockley was. And I certainly didn't know what a sperm bank was. All I understood was that Shockley and the Nobel sperm bank, whoever and whatever they were, had somehow broken my father's code of science. My father was revolted, as though he'd seen the rabbi at our synagogue driving a new Mercedes.

The idea of "Shockley" lodged itself in my ten-year-old brain. He became my symbol of Science Gone Wrong. I pictured Shockley as a cartoon villain in a white lab coat, with a cocky grimace and lunatic, Einstein-like white hair exploding from his head. Decades later, I was astonished when I finally saw a photograph of the real William Shockley, with his mild expression, nerdy glasses, and bald, shiny skull. (I had tricked myself with a visual pun: I had given my "Shockley" a "shock" of hair and the air of a man who'd been electric-shocked.)

My father and I never talked about Shockley or the Nobel sperm bank after that. But twenty years later, at the height of the Internet boom, I encountered Shockley's name again in an article about Silicon Valley. I can barely recall the piece—I think it had something to do with whether Shockley deserved credit as the founder of the Valley. Shockley, the Nobel Prize sperm bank—bells started clanging in my head, in an ominous minor key. I remembered my 1980 conversation with my father. I wondered again what he had been so incensed about. I began reading everything I could find about Shockley and soon learned that he had been one of America's most brilliant scientists, most influential businessmen, and most perplexing racists. After a scientific career of unmatched accomplishment, Shockley had spent the last twenty-five years of his life trying to stop poor people and black people from having children. Thinking Shockley might make a good subject for a biography, I traveled out to Stanford University to dig through his archives.

As I combed through Shockley's papers, I realized that the Nobel sperm bank, not Shockley, was the real riddle. Shockley, I discovered, had been the only *public* donor to the bank—his contribution was one of the last acts of his long and cantankerous career. Shockley's papers were full of articles about the early days of the Nobel Prize sperm bank—a *Playboy* interview with Shockley about it, editorials denouncing Shockley and bank founder Robert Graham, letters from Graham to Shockley. But I found almost nothing from later on. The bank had been launched in 1980 with the immodest goal of changing mankind and reversing evolution. Over the next nineteen years, more than two hundred "genius kids"—as reporters liked to call them—were born from its supersperm. Every few months, some newspaper or magazine or TV network had dispatched a reporter to break open the story of the Nobel Prize sperm bank, to ferret out the donors' identities and learn whether the kids had lived up to their genetic programming. But all the reporters had come back empty-handed.

The Nobel sperm bank finally closed in 1999. Robert Graham and William Shockley were dead. Its records were sealed. And practically nothing was known of the kids, their parents, or their eminent donors. It was the most radical experiment in human genetic engineering in American history, yet no one knew how it had turned out. There was only a beguiling blank: the mystery of the Nobel Prize sperm bank.

I was enthralled with this weird project, this Brave New World, southern California–style. What were the kids like? Had the genius genes created genius babies? Were Repository prodigies now skipping their way through America's best private schools, prepping for Harvard, intent on curing cancer and reinventing physics? Were there lots of little Shockleys out there, hot-wiring the latest Intel chips to work double time? And I still wondered why my father had been so disgusted by it. What was so wrong with the Nobel sperm bank?

I was doing this sperm bank reading in late 2000, right around the birth of my first child. When we were trying to conceive, my wife and I had been thrust into the fertility-industrial complex, that nightmare of ultrasounds, ovulation kits, sperm samples, and hormone injection after hormone injection after hormone injection, our bodies manipulated by doctors as though they were faulty pieces of machinery. Hanna got pregnant before we had to resort to more elaborate treatments, but the fertility industry left me stunned. I talked about it to friends, who recounted stories of their $20,000 IVF treatments or their weeks of shopping at sperm banks—leafing through catalogs, listening to audiotapes, and poring over baby pictures in search of just the right donor. I came across an amazing statistic: one million American children had been born from donor sperm, and 30,000 more were arriving every year. In the two decades since my father had scoffed at the Nobel sperm bank, we had gone from a world where a sperm bank was something exotic and shocking to one where you could practically shop for babies at the mall. How had this happened?

My daughter, Noa's, birth increased my curiosity about the Nobel sperm bank, for a different reason. As Noa slept in her cradle,

I studied her and puzzled over what part of her came from me, what from my wife, and what from God. The shape of her head, the set of her mouth, the wide, round cheeks—those looked to be mine. But what about my interest in the presidents—had she inherited that? My even temper? My profound lack of musical ability? I didn't know yet, and it nagged me.

And the more I thought about her, the more I found myself wondering about the children of the Nobel sperm bank. If I—who had no particular genetic gifts to give—was placing unreasonable genetic expectations on my two-month-old, what must it be like to be the genetic product of a real genius and to have been specifically engineered for brilliance? What was it like when you knew you had "Nobel DNA" powering every nucleus in your body? Did it screw you up? Did it turn your parents into naggy monsters?

The Nobel sperm bank kids, I realized, were messengers from our future. We are on the brink of the age of genetic expectations. Soon—maybe not in five years, but probably in fifty—fertility doctors will be able to identify and manipulate genes for "intelligence" and "beauty." At first, building better babies will be a science, as doctors figure out how to swap genes in order to save kids from terrible diseases. But eventually it will become a consumer movement. Parents will demand the gene treatments not for health reasons, but to make their kids "better." ("Doc, this kid has *got* to play tennis.") Eugenics will be chic again, though not by that name.

The Nobel sperm bank and its children, I realized, had previewed this world to come. I had seen *Gattaca*, a movie about a utopian/dystopian future in which embryos are genetically manipulated to produce enhanced children—a desired piano prodigy is given twelve fingers, for example. The Nobel sperm bank seemed a mini-Gattaca—the same dream of genetic control. Using the most advanced (but still crude) technology of his age—a sperm bank with elite sperm donors—founder Robert Graham had persuaded women they could bear children with the "best" genes available, children who would be smarter, healthier, and happier than nature would have permitted.

I wanted to know if Graham's experiment worked. So in early 2001, I set out to learn the real story of the Nobel sperm bank. Only the sketchiest information about the bank existed, and there was no obvious way to gather more. There were no archives to explore. The names of donors and children were sealed away. But I worked for an Internet magazine, *Slate*, and my editor, Michael Kinsley, was willing to help. We thought we could use the Internet's collaborative power to discover the stories that had never been told. Although we ourselves couldn't find the hidden children, parents, and donors, thanks to the Internet, *they* might be able to find *us*. On February 8, 2001, I published an article outlining the very little I knew about the bank. At the end, I asked anyone who had been involved in it—any child or mom or donor or employee—to contact me.

With their help, I wrote, we could finally unravel the mystery of the Nobel Prize sperm bank.

THE GENIUS FACTORY

CHAPTER 1

THE GENETIC PASSION OF ROBERT GRAHAM

February 29, 1980: The *Los Angeles Times* photograph that introduced Robert Graham and his Nobel Prize sperm bank to the world. *Robert Lachman/*Los Angeles Times

The *Los Angeles Times* headline beckoned like a bulletin from the future: "Sperm Bank Donors All Nobel Winners: Plan Seeks to Enrich Human Gene Pool." It was February 29, 1980, Leap Day—that strange quasi-day seemed right for such an otherworldly story. The article began by describing the sperm bank as "the world's most exclusive men's club," then piled on the weirdness: a reclusive zillionaire . . . a secret cadre of Nobel geniuses . . . the women of Mensa . . . a mysterious, ultramodern fertility technology . . . a sinister experiment to improve the human race. It sounded like something out of a James Bond movie.

The article introduced America to Robert K. Graham—a most unlikely sperm banker. The seventy-four-year-old optometrist, who had made $100 million by inventing shatterproof plastic eyeglasses, was on a mission to collect sperm from Nobel laureates. He was storing the prize seed in an underground bunker on his Escondido, California, estate, and he was distributing it only to women smart enough to qualify for the high-IQ society Mensa. Graham had given his sperm bank a name that had the thud of second-rate science fiction: "The Repository for Germinal Choice."

Graham told *Times* reporter Edwin Chen he had already enlisted three Nobel prize–winning scientists to "deliver" their sperm, and eventually he intended to canvass all the world's Nobel laureates. So far, Graham said, two dozen Mensa women had contacted him—he had told the *Mensa Bulletin* about the bank a few months earlier—and he had shipped frozen Nobel sperm to three of them.

The sperm bank was not a prank, Graham insisted to Chen, and not a rich man's folly. Graham said he was trying to save mankind from genetic catastrophe. In modern America, the millionaire complained, cradle-to-grave social welfare programs paid incompetents and imbeciles to reproduce. As a result, "retrograde humans" were swamping the intelligent minority. This "dysgenic" crisis would soon cause the evolutionary regression of mankind, as well as global communism. How could we save ourselves? Graham had the answer. Our best specimens—and "specimens" is just the kind of word Graham would use to describe people—must have more children. His Nobel Prize sperm bank would father a cadre of leaders, scientists, and politicians who would help reverse the genetic decline. Graham was not charging his customers or paying his donors. He and his Nobelists were making a gift of the genius genes, a lifesaving present to a dying world. Graham promised to study the children of his supersperm, tracking their development, achievements, and IQ. He would publish his findings in scientific journals, vindicating his extraordinary semen and his experiment.

Graham outlined his ideas to Chen with an unapologetic bluntness. "The principles of this may not be popular, but they are sound," he said. "So far, we've refused to apply to humans what we already know and apply to animals and plants."

Graham gave Chen a tour of the bank, such as it was. Graham owned ten beautiful acres in Escondido, a thriving town half an hour northeast of San Diego. In Graham's prizewinning garden, in the shadow of the American flag that Graham flew over his property, sat a concrete bunker the size of a modest bathroom. The bunker was a few feet underground and slightly dank. It had once been a pump house; Graham had converted it into a small laboratory. Its prize equipment was a lead-sheathed, waist-high vat of liquid nitrogen. Graham opened the vat and showed Chen what he claimed was the Nobel sperm, a few dozen ampules frozen at 196 degrees below zero centigrade.

Graham wouldn't disclose the names of his three Nobel donors, so Chen wasn't sure if Graham was an honest eccentric or a con artist. After all, Nobel laureates' sperm looked like anyone else's; Graham's vials could just as well have contained seed from his gardener. Chen called every Nobelist he could find in California and asked if he had made a sperm deposit. One after another said no. A dozen had never heard of Graham's project. Another ten admitted that Graham had contacted them but said they had turned him down. Most were scornful. "It's pretty silly," Medicine Laureate Max Delbrück told Chen. Chemistry and Peace Laureate Linus Pauling said he had nixed Graham: "The old-fashioned way is still best." Medicine Nobelist Renato Dulbecco, who hadn't been contacted, burst out laughing at the idea, "Oh come, come. This is fantastic. . . . It's too late for me. I was vasectomized long ago."

Finally Chen reached Stanford's William Shockley, whose invention of the transistor had won him the 1956 Physics Nobel. Chen asked his question. After a long, long pause, Shockley said, "Yes, I'm one of them. This is a remarkable attempt, and I am thoroughly in sympathy with this sort of an approach. Everyone talks about it, but by God, Graham is doing something about it." Chen had his story. The Nobel sperm bank was for real.

A remarkable photo of Graham illustrated the *Times* article. At first glance, nothing seemed out of the ordinary. It showed a handsome elderly man with perfect posture and a formal bearing. Graham wore a camel's-hair blazer and a perfectly knotted polka-dot tie. His crown of thick gray hair was Brylcreemed to a high shine. It looked like a photograph from a midcentury corporate annual report: Our CEO at work. But then you noticed Graham's eyebrows were jauntily cocked. And his hands were sheathed in weird, massively thick oversized mitts. And he was standing in a laboratory, not a boardroom, with a microscope perched on a shelf behind him and a huge metal drum at his feet. The caption read: "Robert K. Graham in his underground chamber. At lower left, the sperm repository, made of thick lead to shield it from radiation." The picture was a mesmerizing conflation of future and past, the 1950s businessman and his twenty-first-century project. All at the same time it exuded optimism, pragmatism, malevolence, and goofiness.

Chen's *Los Angeles Times* article provoked an international sensation. Chen has been a journalist for more than thirty years on some of the most important beats in the world (he now covers the White House for the *Times*), and he says he has never experienced anything like the frenzy of February 29. Journalists called him by the dozen; so did desperate women hoping to score genius sperm. Graham, too, was inundated with media requests. Reporters from all over the country wanted to see his magic vials and quiz him about his intentions. Graham's subterranean chamber became the ground zero of the future. The press immediately gave the "Repository" a much flashier nickname: the "Nobel Prize Sperm Bank." The nickname pleased Graham, who started using it himself.

On March 2, Graham held a press conference in his backyard. He spent most of the session rebutting accusations of sexism, racism, elitism, white supremacism, and Nazism. Yes, Graham conceded, all his donors were white, and it was true that he gave sperm only to married women, not lesbians or single women. But he was no Nazi. "I don't know much about Hitler and his vision. I don't see a parallel. We aren't thinking of a super-race. We are thinking in terms of a few more creative, intelligent people who otherwise might not be born."

But in giddier moments, Graham was dreaming bigger than just "a few more creative, intelligent people." His little Repository was a "pilot project," he said. Soon it would seed similar banks around the world. Every city would have its own genius sperm bank. There wouldn't be just a few superkids, but thousands of them. Given enough time, Graham mused, genius sperm banks might help "stimulate [man's] ascent toward a new level of being, of which his present organic status may be only the crude beginning." His Repository, Graham hoped, might one day give birth to mankind's "secular savior."

Other people could have conceived the idea of a Nobel sperm bank, but no one except Robert Graham could have conceived it *and* made it a reality. Graham had the right-wing politics of a self-made millionaire, the relentless inquisitiveness of an inventor, the can-do spirit of an entrepreneur, and the moxie of a salesman. Together, these qualities made him confident that his sperm bank was the right idea, rich enough to fund it, and certain that he could market it to a skeptical public. It was also no accident that Graham was in southern California, the ground zero of American libertarianism. In 1980, California culture was a clash between freethinking futurism (new-ager Jerry Brown, "Governor Moonbeam," was in the middle of his second term) and hard-right political conservatism (former governor Ronald Reagan had swept the New Hampshire primary just three days before the *Times* article). In Robert Graham—and perhaps *only* in Robert Graham—these alien theologies intersected. His sperm bank sought to harness scientific libertarianism and dreamy futurism, and put them in the service of rigid social control.

Here's my favorite Robert Graham story.

In the early 1970s, when he had tired of running his eyeglasses company but wasn't yet collecting Nobel sperm, Robert Graham tried to start a country. He thought an island would be best. Graham instructed George Michel, a vice president at his firm, Armorlite, to locate an island that Graham could buy and flag as a sovereign—or at least semisovereign—nation. Graham instructed Michel that the island should be at least five miles wide and fifteen miles long.

Michel hired several Los Angeles real estate agents, and they eventually located four or five promising candidates, mostly small islands in the Atlantic Ocean that Great Britain might surrender for the right price. Graham was thrilled. Next, he assigned Michel and several Armorlite colleagues to design the island's living and working quarters. Graham decreed that the island had to be completely self-sufficient and that no cars would be permitted on it. Michel drew blueprints for prefabricated living saucers that could be stacked on land or in the sea. He designed a futuristic sewage system, greenhouses, and food factories. His masterpiece, Michel recalls fondly, was a vacuum-tube-driven transportation system, in which gyroscopically balanced pods would zip passengers from one part of the island to another.

Graham's island wasn't the usual kind of millionaire's ego trip. Graham didn't aspire to rule his kingdom. He lived to play handmaiden to great men—men he thought were better than he was. Graham intended to create an elite research colony. Graham would invite the world's best practical scientists to the island, offer them lavish living conditions and the fanciest laboratories money could buy, and let them start inventing. Grahamland would support itself: when scientists produced something valuable, they and the colony would share the royalties. The inventors would get rich, and Grahamland would prosper. Graham was convinced that scientists would flock to his island, because he was sure they wanted what he wanted: an escape from the morons, weaklings, and imbeciles who increasingly dominated the rest of the world. Science would be Grahamland's god and its law. It would be a rational empire, Graham's own private Atlas Shrugged.

Grahamland never progressed beyond the planning stages. Michel quit Armorlite in a stock dispute. Graham got distracted and never managed to buy the island. But the private nation was pure, distilled essence of Robert Graham: the entrepreneurial vigor; the cockamamie grandeur; the unshakable faith in practical science; the contempt for the pig-ignorant, lazy masses; and the infatuation with finding—and *claiming*—the world's best men.

Robert Graham was born on June 9, 1906, in Harbor Springs, Michigan. When he was a rich old man, Graham liked to tell stories that made it sound as if he'd grown up on the frontier—kerosene lanterns instead of electricity, hauling water up the road for the Saturday-night bath. This was bogus nostalgia. Though Harbor Springs was small (only 1,500 residents) and rural, there was no pioneer hardship. Harbor Springs was a resort, the summer playground of midwestern royalty, and it was enjoying its heyday as Robert was growing up in the 1910s and '20s. The town sat on Little Traverse Bay, a gorgeous inlet of Lake Michigan at the northern tip of the state. Harbor Springs famously had the cleanest air in America: the west winds racing over the lake stripped the air of pollen and dust. Hay fever sufferers made Harbor Springs a summer refuge in the late nineteenth century. Thanks to railroads and a ferry, the rich soon followed for the beautiful harbor and the long summer nights. The Harbor Springs summer census was a who's who of American business: Cincinnati's Gambles (Procter & Gamble), Louisville's Reynoldses (aluminum), Illinois' Pullmans (trains), Michigan's Upjohns (pharmaceuticals), and many other names from the fronts of supermarket packages and the backs of automobiles. They built "cottages"—Newport-like mansions—along the town's glorious beaches. And they got together to establish Michigan's most exclusive country clubs, purchasing huge tracts of land, fencing them off, and laying out the first golf courses in the state.

Robert Graham was born into the local gentry—the respectable year-rounders who were acknowledged by the summer folk but not of them. His father, Frank Graham, was the local dentist and prospered by treating both locals and tourists. Frank had graduated first in his class at the University of Michigan dental school, married Fern Klark, and settled in Harbor Springs in 1903. He built himself a fine shingled Victorian house on East Bluff Drive, the street where the richest townies lived. Fern Graham was a gracious, gentle woman, but Frank was chilly and formal. When he took a walk on the town beach, he wore a coat and tie. Frank was a clever man, however, an inveterate tinkerer. He invented a new carburetor for boat engines and designed a collapsible keel for sailboats. After the *Titanic* sank, Frank spent years trying to build a better lifeboat.

The oldest of four children, Robert inherited his mother's grace and his father's inventiveness and formality. But he may have inherited even more from Harbor Springs. Growing up in the resort town instilled in Robert Graham a lifelong obsession with the rich and the great. (This is a man who titled the longest chapter in his autobiography "Princes and Princesses I Have Known.") In summer, Graham caddied at Harbor Springs' two private golf courses, Harbor Point and Wequetonsing. Graham caddied to earn pocket money— 75 cents for eighteen holes—but also to spend time around powerful men. The respectful adulation he would perfect as a genius sperm banker—he learned that on the golf courses of Harbor Springs. Some eighty years later, Graham wrote about caddying in his memoir: "I know of no other situation in which a boy can be in the company of leading and outstanding individuals, hours at a time. He can learn some of their ways of thinking and talking, their matters of concern and some of their foibles." (Graham carried the bags of famed baseball commissioner Judge Kenesaw Mountain Landis, among others.) Caddying prodded Graham's ambition in another way: it made him greedy. The Graham family was plenty prosperous, but he was a townie, a second-class citizen, practically a ragamuffin compared to Harbor Springs' majestic summer migrants. Like many middle-class kids who spend their lives around the rich, Graham smelled money and developed an appetite for it. "I saw these wealthy summer people enjoying themselves at leisure and concluded that wealthy was the way to be."

Harbor Springs is also where young Robert learned his first, unfortunate lessons about race. Graham's ancestors were fairly recent arrivals in America—there were Welshmen and Czechs in the near past—but they were very white and very Protestant. These traits were virtually requirements for living in Harbor Springs, a town that was as prejudiced as you'd expect for Michigan in the 1920s. This, after all, was when Henry Ford was at the height of his influence and his anti-Semitism. Jews and blacks were excluded from Harbor Springs' clubs, of course—not that there were any around, except for the black servants of some summer folk. The many Indians who lived around Harbor Springs were second-class citizens, mostly confined to jobs in manual labor. In Harbor Springs, Graham developed the discomfort with nonwhites that he would never lose.

By the time he graduated from Harbor Springs High School in 1924, Graham had acquired a distinctive bantam charm that he would carry till his death. He was not a tall man—five feet, eight inches on a good day—but he carried himself like one. He had the posture of a Prussian Army colonel, and his head was huge for his

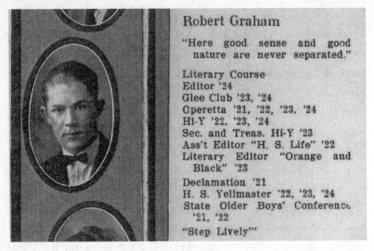

Robert Graham

"Here good sense and good nature are never separated."

Literary Course
Editor '24
Glee Club '23, '24
Operetta '21, '22, '23, '24
Hi-Y '22, '23, '24
Sec. and Treas. Hi-Y '23
Ass't Editor "H. S. Life" '22
Literary Editor "Orange and Black" '23
Declamation '21
H. S. Yellmaster '22, '23, '24
State Older Boys' Conference '21, '22
"Step Lively'"

Robert Graham in the Harbor Springs High School yearbook, 1924. *Courtesy of the Harbor Springs Library*

body. Graham was vain, and he had much to be vain about: His chest was broad from swimming, his legs strong from running. He had jug ears, eyebrows so bushy they looked fake, and a vast chin— but these aggressive features combined in a fortunate way. His hair, thick and brown with a widow's peak, was slicked back in the fashion of the day. Many girls thought him gorgeous, and he knew it. His classmates voted him the best-looking boy in the class of 1924. They liked him, too. Next to his picture in the Harbor Springs High School yearbook, it said, "Here good sense and good nature are never separated."

At eighteen, Graham headed off to the University of Michigan in Ann Arbor, intent on shedding his bourgeois roots. Frank Graham had urged his eldest son to follow him into dentistry, but Robert loathed the idea of "fooling around in people's mouths." He wanted to do something more ambitious. He set out to become the next Enrico Caruso. His voice had dazzled audiences in Harbor Springs, and he believed he could be a star. He spent eight years studying music at the University of Michigan and Michigan State University. He sang leads in student operas and twice soloed at New York's Metropolitan Opera House during university tours.

In 1932, at the pit of the Great Depression, he graduated and moved to New York City to be discovered. He wasn't. He blew an audition at Radio City Music Hall. With the brutal rationality that would eventually make him a good businessman, he recognized that no amount of teaching would ever make him a Caruso: his voice was too erratic. He returned abjectly home to Michigan. Very quickly, he blotted his musical career from his memory. It had been, he said, a "waste." When he wrote his memoir sixty years later, he scarcely mentioned it and left out entirely the fact that he had married another singer and had children with her. All that Graham let himself remember from his singing years was what he *always* remembered: his brushes with famous men. How his uncle, the celebrated architect Ernest Graham (the Wrigley Building, Equitable Building, Flatiron Building), had paid for his New York music lessons. How he had befriended Arnold Gingrich, later the founding editor of *Esquire*.* How he had spent a weekend at Frank Lloyd Wright's house, where he had been awed to see Wright so engaged in "exalted discourse" that the architect didn't notice he was drooling egg yolk all over his shirt and tie. In a poor man, Graham would have considered dripping egg yolk the sign of a slovenly mind. But in Wright, it meant genius.

Graham remained determined to get rich, but he needed a new path. He was blessed with a clear-eyed view of himself. "[I] have no great gifts, but no great weaknesses, either." He knew he was disciplined: He never drank alcohol or coffee, never smoked, never gambled. He solved problems quickly, and his hands were as agile as his

---

* Graham took credit for the magazine's name. At the time it launched, he was facetiously addressing his letters to "Arnold Gingrich, Esquire." When Gingrich was brainstorming a name for the magazine, one of Graham's letters arrived, and *Esquire* was born.

mind. He loved hard work and believed in its moral virtue. With all this in mind, Graham settled on a second career: optometry. It was an odd but inspired choice. Though deeply unglamorous, optometry was a profession of gadgets—not very good gadgets. Graham relished the challenge of trying to improve eyeglasses and the tools that made them. He earned an optical degree from Ohio State in a mere eighteen months—inventing a new kind of lens along the way—and landed a coveted job at Bausch & Lomb immediately after graduation. When America entered the Second World War, Graham was a father in his mid-thirties. He spent the war figuring out how to use the optical technology in captured German equipment to improve American artillery scopes and binoculars.

When the war ended, Graham was working for the optical giant Univis. It was a drag. Graham was a salesman, and he was good at it—gracious, elegant, smart—but his heart wasn't in it: he lacked the salesman's profligate bonhomie; he didn't have the patience to explain things to people he thought were stupid. Graham liked the *tinkering* of optometry, not the salesmanship. So Graham threw himself at the profession's number one problem: Why were eyeglasses so bad? Lenses were still made of glass, which meant they were fragile and dangerous. Thousands of Americans suffered eye injuries every year when their spectacles shattered.

Graham saw the future, and it was . . . plastics. Despite decades of attempts, no one had been able to manufacture a plastic lens that was as reliable and scratch resistant as glass. Graham thought he could. In 1947, when Univis refused to dedicate itself to plastic lens research, Graham quit, recruited a partner, and poured all his money into starting a new company, which they called Armorlite. Graham moved to southern California, the red-hot center of the postwar industrial economy, and tried to make plastic eyeglasses. He failed and failed and failed. After a fiasco using Plexiglas, Graham began to experiment with a little-known plastic called CR-39. It had been used to make B-17 fuel tanks during the war.

CR-39 was a disaster, too. It shattered the lens molds, and it shrank too much as it dried. But Graham persisted with it and perfected CR-39 lenses at the end of 1947. Armorlite's lenses revolutionized the optical business. In the 1950s, Armorlite thrived but still served a niche market. Then fashion came to Graham's aid. Large lenses were the vogue of the 1960s, and they could be made only of lightweight plastic. Armorlite boomed. Graham employed five hundred workers at his Pasadena factory. He marketed his product aggressively and was a great showman: When he gave a speech, he would yank off his Armorlite spectacles, fling them into the air, let them fall as the audience gasped, and then pick them up, unscathed. Graham kept on tinkering—he helped perfect contact lenses, developed the first antireflective coating for plastic lenses, and manufactured the first UV-protective lenses, among other inventions. He was a hero in his small corner of American business. Optical societies rained medals down on him. The National Eye Research Foundation dubbed him "The Man Who Made It Safe to Wear Glasses." He also became incredibly rich. During Armorlite's lean years, Graham took his salary in stock; by the time Armorlite struck gold, he controlled nearly the entire company. Graham risked all on Armorlite and made it back thousands of times over.

But Graham was dissatisfied. His personal life was messy. Graham had divorced his first wife after she had borne him three children, then played the field with a sportsman's relish. He was an incorrigible flirt, and his sharp good looks, dapper dress, and impeccable manners helped his cause. He remarried, unhappily, to a woman his brother Tom described as an "alcoholic showgirl." That miserable union produced two more children, but it was headed toward divorce when his wife swallowed an overdose of sleeping pills and died. He wiped this second wife from his history books, too. (Graham possessed the great American gift of amnesia. He forgot nasty parts of his past as if he were erasing a chalkboard.) With a lack of awareness that would be funny if it weren't sad, he described the second marriage in his memoir with a single sentence: "I had recently concluded an embittering marriage and swore never to put my neck into that noose again." He fathered another child—out of wedlock, according to his brother—and then found wife number three in 1960. Marta Ve Everton, an ophthalmologist twenty-one years his junior, was whip smart, elegant, religious, and altruistic. She was the great love of Graham's life. She bore him two children, bringing his total to eight.

Graham had an ambivalent relationship with his brood. He liked the idea of family in theory but bungled it in practice. Like his own father, he was emotionally distant with his kids. His three daughters thrived, especially Marta Ve's two girls. But three of his boys found serious trouble. One apparently killed himself. Another suffered a traumatic head injury as a boy, never quite recovered, and died in middle age after a difficult life. A third moved to the Pacific Northwest and cut his ties with the family. Graham seemed ashamed of some of his sons; he would sometimes avoid introducing them to his friends.

Graham's success in a too-narrow field, his huge, almost-but-not-quite-happy family, his fascination with the rich and famous: in the late 1950s, all these helped inspire the passion that would define the rest of his life. Graham came to believe—more strongly than he believed anything—that society was doomed unless smart people had more children. He vowed to help them do it.

Graham's obsession began with a mistake. Graham's childhood idol had been an inventor named Ephraim Shay. Shay had designed the "Shay locomotive"—a powerfully geared steam train that could climb steep hills. Mining and logging firms bought Shay trains by the hundreds. He made a fortune and retired to Harbor Springs in the late nineteenth century. He was the town's most celebrated resident, famed for his fertile mind and generous heart. He engineered Harbor Springs' water supply. He built experimental boats that he docked in the town harbor. In winter, he hammered together hundreds of sleds for the town's children, including young Robert Graham. When Shay died in 1916, it hit ten-year-old Robert hard. He believed that Shay had died childless. The inventor's barrenness lodged in Graham's head and eventually goaded him to act. As an adult, Graham would write, "God had planted some of His best seed in our town, but it had died out. They still name streets and schools for Ephraim Shay. The great bronze tablet which recounts his accomplishments still stands. But the genes which determined his extraordinary nature have died out. Ever since, the extinction of exceptionally valuable human genes has been a concern of mine."

In fact, Graham was wrong about Shay. The inventor's "seed" was alive and well and spreading all over America. Shay had fathered a son before moving to Harbor Springs. In 2000, three years after Graham's death, about 160 of Shay's descendents, all carrying his "exceptionally valuable" genes, celebrated their ancestor at a reunion in Harbor Springs.

Shay was Graham's chief inspiration, but not his only one. All around him, Graham glumly observed the triumph of dullards over brains. Graham sold contact lenses to pro football players, and he was repulsed at how women flung themselves at the mountainous morons. Graham would sometimes eat lunch in the cafeteria of his Pasadena factory—sometimes, but not often, because his employees irritated Graham too much. He thought they didn't want to improve themselves or work harder. All they cared about was milking the government for more benefits. These indistinct resentments clarified themselves in Graham's mind. He had no religion, but he found a faith in eugenics. He became fixated on the idea that the world needed more intelligent people because the idiots were multiplying too fast.

It was not surprising that Graham grew fascinated with genetic degradation when he did, as the 1950s turned the corner into the '60s. The late 1950s had marked the zenith of men like Graham. In the Sputnik-era scientific-industrial complex, technical businessmen were kings. White men just like Graham—intelligent, arrogant, scientific, and self-assured—dominated 1950s America. (Their rationalist ethos didn't merely pervade business and government, it also spilled over into other, less obviously scientific aspects of human life. Alfred Kinsey's pioneering studies, for example, helped popularize the notion of sex as a mechanical act, separable from human emotion. Very little separated Kinsey's scientific sex from Graham's scientific breeding.) As a titan of industry and prize inventor, Graham felt he had the right, even the obligation, to impose his eugenic ideas on the idiotic masses. Graham's genetic dread also reflected his fear of the societal change that he sensed was coming. Graham began worrying about the intellectual decline of Americans at the very instant Americans started to decide they didn't want to listen to men like Graham. The civil rights and women's movements were overthrowing the white male order. The demand for Wise Men was withering.

So at this nervous moment, with the Wise Men still clinging to power, Graham wrote a book to sound the alarm. Part pseudoanthropology, part evolutionary biology, all polemic, Graham's *The Future of Man* throbbed with panic: Act now or humanity will die! The thrust of *The Future of Man* was that prosperity had ruined mankind, because it had reversed human evolution. Graham, undeterred by the fact that more people were living longer, healthier, and richer lives than at any point in history, concluded that man had peaked 15,000 years ago, in the good old days of the Cro-Magnons. These "scourging gods," as Graham called them lustfully, had been brilliant and mighty because nature was so ruthless. Only the greatest Cro-Magnons had survived to pass on their glorious genes. But then, the tragedy of civilization! The agricultural revolution had softened man and allowed weaker specimens to breed. Since intelligence was 50 to 90 percent inherited, according to Graham, mankind got stupider as these lesser men multiplied. Natural selection waned. After thousands of years of such regression, half the human population was "what might be described as dull." Graham believed that the spread of half-wits explained the rise of communism, a political ideology that squashed brilliance and rewarded mediocrity.

Graham anguished that the few smart people who remained were cooperating in their own extinction by using birth control, an "almost wholly pernicious" invention. The refusal to reproduce was "nature's unforgivable sin," Graham wrote.

> The disappearance of genes for high intelligence is a defeat for the uniqueness of man, an erosion of the essence of the human condition. The childlessness of an Isaac Newton or a George Washington, the extinction of the Lincoln family, the spinsterhood of the brightest girl in the class, are great biological tragedies. As a result, mankind is deprived of some of that essential quality which separates him from the apes.

But a remedy was at hand. Just a few more smart people, and we could fend off the idiotic hordes. "Ten men of high intelligence can be more effective than 1000 morons." Mankind, Graham proclaimed, could seize control of evolution through "intelligent selection."

Ever practical, Graham ended *The Future of Man* with a how-to guide for saving humanity. Most of his proposals were mundane — government support for married graduate students, housing projects designed for large families, corporate sponsorship of employees with children.

But Graham saved his one big idea for last: "germinal repositories"—or, as they would be come to be known, sperm banks. Women would be artificially inseminated with sperm collected from the world's smartest men: "Consider what it would mean to scientific progress if another 20 or more children of Lord Rutherford or Louis Pasteur could have been brought into the world. . . . Consider the gains to society if this new technique had been available to engender additional sons of Thomas Edison." The number of geniuses in the population, Graham declared, would "increase exponentially."

Graham was a man out of time. He didn't realize that he had arrived at his views seventy-five years late. His ideas were so old they were new again.

CHAPTER 2

MANUFACTURING GENIUS

Fitter Family contest winner, Eastern States Exposition, 1925. *Courtesy of the American Philosophical Society*

When I tell someone I'm writing a book about a Nobel Prize sperm bank, this is the usual response: First a quizzical "It's a novel?" Second—after I shake my head "no"—a laughing double-take. "You're kidding. That's a joke, right?" At a distance of twenty years, Robert Graham's sperm bank does seem like nothing but a giggle. Yet there was a time when it could be taken not merely seriously but as the most serious thing in the world. There was a time when celebrated men would risk their reputations on such an idea.

Robert Graham had come late to the eugenics craze that had

gripped the United States and Britain for fifty years from the late nineteenth century until the 1930s. Eugenics was the confluence of three rivers of Anglo-American thought: late-eighteenth-century theories about overpopulation, late-nineteenth-century Darwinism, and early-twentieth-century racial paranoia. The gloomy eighteenth-century philosopher Thomas Malthus outlined the first key ideas of what would become eugenics. Fearing overpopulation, Malthus concluded that the poor's misery must be ordained by Mother Nature. Their suffering and early death were good, Malthus said, because it prevented them from spreading their innate weakness. Any effort to ease their lot, with, say, a minimum wage, would increase misery in the long run. A century later, Social Darwinism gave a scientific framework to Malthus's instincts. If Darwin's "survival of the fittest" ruled birds, surely it ruled mankind, too. Mother Nature intended to cull the worst humans before they could breed, while the best of us were obliged to go forth and multiply.

It was a cousin of Charles Darwin, Francis Galton, who transformed these anxious theories into a new "science" of eugenics. Galton, an eccentric and adventurer, was obsessed with measuring anything that could be measured. He amassed weather data and composed Britain's first weather maps. He pioneered the use of fingerprinting. And in the 1860s, he set out to measure human achievement. Galton's 1869 book, *Hereditary Genius*, counted and classified Britain's most accomplished men and showed that they were very often related to one another. Successful fathers had successful sons. This, Galton claimed, proved that God-given abilities were passed from one generation to the next. (It did not concern Galton that in Victorian England, advantages of birth, wealth, and education might have given the sons of famous men a career boost.)

Galton named his new science "eugenics," an invented word based on the Greek for "well born." For Galton, the goal of eugenics was to increase genius. The best people must be prodded to reproduce, because their children's natural gifts would improve Britain.

Galton's acolytes, however, immediately focused on the dark reverse of his theory: If the rich are rich because they are endowed with natural abilities, the poor must be poor because they are endowed with natural inabilities. Why were there so many poor criminals, imbeciles, drunks, epileptics, and morons? Why were the poor so shiftless? Why were the poor so . . . poor? Because they were *naturally* weak. "The taint is in the blood."

The British talked plenty about eugenics, but it was can-do Americans who converted Galton's theory into dismal practice. Americans adopted eugenics with a convert's zeal. In the United States, eugenics quickly merged with racial anxiety: blacks and immigrants from southern and eastern Europe—that is, Negroes, Jews, Papists—were threatening to overwhelm America's white, Protestant, northern European elite.

Eugenics was a way to fight back. With vigorous American entrepreneurship, eugenicists took Galton's philosophy, spiced it up with a dollop of Mendelian genetics, and turned it into an aggressive, impolite, and wildly successful national crusade to preserve the American "germ plasm." In 1910, Charles Davenport opened the Eugenics Record Office in Cold Spring Harbor, Long Island. Rockefeller, Carnegie, Harriman—the richest families in America supplied the funds for it. Davenport and his assistants scoured America in search of the "unfit." They hunted for albinos, the Amish, epileptics, mental patients, and criminals and cataloged their supposed genetic weaknesses on note cards. (Eventually, the ERO would collect 750,000 of the cards.) Anthropologists wrote case studies of depraved families. The Jukeses of New York were the subject of not one but two books detailing the family's array of criminals, lunatics, and imbeciles. Eugenicists gave new, allegedly scientific intelligence tests to immigrants at Ellis Island and discovered that 80 percent of them qualified as "feeble-minded."

Americans—that is, white upper- and middle-class Americans—took to eugenics like a cult. University presidents, academics, congressmen, businessmen, and good society everywhere embraced the creed. Eugenics was proselytized by everyone from Ivy League professors to the KKK. By the late 1920s, 20,000 college students a year were taking eugenics courses. Eugenics assumed the trappings of a religion: Eugenicists proposed a "Decalogue of Science"—a revised, eugenic Ten Commandments. The American Eugenics Society sponsored eugenics sermon contests and issued a Eugenics Catechism:

Q: What is the most precious thing in the world?
A: The human germ plasm.

American eugenics leaders weren't satisfied with merely identifying the unfit. *Neutralizing* them was the goal. The eugenicists persuaded legislatures to write laws preventing "mental defectives" from marrying. They won passage of the 1924 Immigration Act, which choked off the flow of immigrants from southern and eastern Europe. The eugenicists flirted with euthanasia: in 1915, Chicago doctor Harry Haiselden was lionized when he refused to operate on a "defective" newborn, who quickly died. *The Black Stork*, a movie about (and starring) Haiselden, became a national sensation, playing for a decade.

The American eugenicists' most important cause was sterilization. How they longed to cut! They thought practically everyone should get the knife: the "feebleminded," alcoholics, epileptics, paupers, criminals, the insane, the weak, the deformed, the blind, the deaf, and the mute—and their extended families. Of course, most of the purportedly genetic ailments developed by eugenicists were not, in fact, genetic in origin. And even if they had been genetic, sterilization would have been a hopelessly bad cure for them. It would have taken literally thousands of generations of mass sterilization to significantly reduce the incidence of genetic diseases. But eugenicists didn't stop to do the math.

Once surgical vasectomy was perfected at the turn of the twentieth century, the sterilizers got to work. In 1907, Indiana passed the first law allowing the forced sterilization of the feebleminded. By 1917, fifteen states had legalized eugenic sterilization. By the 1930s, a majority of states mandated the sterilization of the unfit. Daniel Kevles's superb history *In the Name of Eugenics* quotes Pennsylvania governor Samuel Pennypacker, the rare public official who opposed a sterilization law. When a political crowd got rowdy on him, Pennypacker retorted, "Gentlemen, gentlemen! You forget you owe me a vote of thanks. Didn't I veto the bill for the castration of idiots?"

Despite state laws, sterilization remained constitutionally murky. But to its opponents, it was a cruel and unusual punishment. To supporters it was an essential public health measure. In 1927, eugenicists finally pushed a case to the U.S. Supreme Court. *Buck v. Bell* concerned eighteen-year-old Carrie Buck, who had been committed to Virginia's Colony for Epileptics and Feebleminded. The state, having declared her an imbecile, was proposing to sterilize her. Under Virginia law, three generations of a family had to be feebleminded before sterilization was permitted. Carrie's mother, Emma, also held at the colony, was classified as feebleminded. Carrie made two defective generations. The state then set out to prove that Carrie's *seven-month-old* daughter was also a moron, thus establishing a third generation. A Red Cross worker testified that the infant had "a look" that was "not quite normal" about her. That was enough for the Court, which voted 8–1 to sterilize Carrie Buck. Justice Oliver Wendell Holmes wrote the majority decision, which likened sterilization to castration. Holmes's words still sting: "It is better for all the world, if instead of waiting to execute degenerate offspring for crime, or to let them starve for their imbecility, society can prevent those who are manifestly unfit from continuing their kind. . . . Three generations of imbeciles are enough."

After *Buck v. Bell*, sterilization became common. Virginia, one of the most enthusiastic states, made raids into Appalachia, rounded up families of "misfit" hillbillies and dragged them down to the asylum operating room. By the end of the 1930s, more than 35,000 Americans had been forced under the knife. Another 25,000 were sterilized before the practice finally petered out in the 1960s. The most unfortunate victims of the mills were children who "voluntarily" submitted to sterilization. In *War Against the Weak*, Edwin Black quotes a transcript of one such child's "consent":

> DOCTOR: Do you like movies?
> PATIENT: Yes, sir.
> DOCTOR: Do you like cartoons?
> PATIENT: Yes, sir.
> DOCTOR: You don't mind being operated on, do you?
> PATIENT: No, sir.
> DOCTOR: Then you can go ahead.

Americans cheerfully exported their bloody-minded eugenic ideas to the world. German eugenicists were particularly captivated by the American notion of Nordic supremacy. Germans published textbooks based on American ideas, and Adolf Hitler read them. He wrote fan letters to leading American eugenicists, telling Madison Grant, for example, that his book *The Passing of the Great Race* was his "bible."

The purported science of American eugenics helped Hitler medicalize and sanitize his hatred, making it palatable for a mass audience. When Hitler took power, he imposed draconian sterilization laws of the sort that his American teachers had only dreamed of. In only three years, the Nazis sterilized 225,000 Germans. When the war arrived, sterilization degenerated into "mercy killing"—the outright murder of tens of thousands of asylum residents. The eugenic murders were the prelude to, and inspiration for, the Holocaust. Nazi eugenic enthusiasm flourished even in the death camps, as

Josef Mengele and his ilk conducted barbaric experiments on twins and other unfortunates in the name of gene science.

The Great Depression had fostered a skepticism about eugenics in the United States. *Those Carnegies and Rockefellers who ruined the nation—we're supposed to believe they're our genetic superiors?* Hitler's crimes sealed the case against eugenics. Disgraced by the war, the sterilizers and race theorists shrank from public attention.

Even as America had been sterilizing its citizens, it had also been flirting with a more innocuous, almost goofy, form of eugenics. Since Galton, eugenics had followed two tracks. "Negative" eugenics— with all the grimness the name suggested—stopped marriages and compelled sterilization to stop the "unfit" from breeding. Negative eugenics was state-sponsored and brutal. But "positive" eugenics took a milder approach. Like Galton himself, the positive eugenicists didn't worry much about punishing the unfit; instead they sought to increase the number of outstanding people. Philosophically, positive and negative eugenics were identical: both embraced the fiction that white Protestants were genetically superior to everyone else; both were founded on a terror of immigrants and blacks; both held that the eugenics crisis was man's greatest challenge. And many eugenicists believed in both positive and negative tactics. But when it came to action, positive eugenics was essentially harmless.

While negative eugenics was goose-stepping toward the gas chamber, the positive eugenicists were embarking on more innocent projects. Popular textbooks instructed young women in their obligation to marry eugenically fit men. Bright young things were warned against the trendy new practice of contraception, because America needed more good, healthy babies. Teddy Roosevelt, an enthusiast for positive eugenics, declared that "the prime duty, the inescapable duty, of the *good* citizen of the right type is to leave his or her blood behind him in the world." Six children, TR averred, was the right number for preventing "race suicide," four the bare minimum.

In classic American style, the positive eugenicists turned the crusade into a competition. In 1920, the Kansas Free Fair hosted the first "Fitter Family Contest." Twenty families entered, and trained eugenicists gave them psychiatric evaluations and intelligence tests. The winning family was paraded around the grounds like prize cattle. Soon the American Eugenics Society was sponsoring "human stock" competitions at fairs all around the country. Winners were awarded a medal on which was inscribed: "Yea, I have a goodly heritage."

The most durable idea in positive eugenics was the dream of turning outstanding men into reproductive machines. The roots of this idea were ancient. In her excellent book *Quest for Perfection: The Drive to Breed Better Human Beings,* Gina Maranto writes that in fifth-century B.C. Sparta, a husband could dragoon one of the city's finest young men to impregnate his wife in order to produce "well-born children." Socrates advised in *The Republic* that the state should breed its citizens like horses, assigning the best men and women to reproduce.

When eugenics took hold in Europe in the 1880s, Maranto notes, the notion of putting the finest men out to stud was revived. Count Georges Vacher de Lapouge proposed that "a very small number of males of absolute perfection" be used to father all children. Lapouge and other eugenicists were mesmerized by the male reproductive capacity. A woman could bear only one child in a year, but a man might father hundreds every day. It was enough to make eugenicists giddy: Why not a whole nation of Pasteurs or Franklins? This mathematical conception of male fertility combined with a mechanical conception of female fertility: women were "receptacles" for children. The product was a potent, if rather icy, vision of the future: the genius factory.

The modern paradigm of the genius factory was laid out by J.B.S. Haldane in a weird little tract entitled *Daedalus.* Haldane, a Brit and Marxist, was one of the twentieth century's great scientists;

he forged the connection between Mendelian genetics and evolution. *Daedalus*, published in 1923, was a scientific prophecy that looked back from the year 2073. Haldane predicted that 1923's primitive eugenics would develop into sophisticated "ectogenesis": eventually, children would be bred in test tubes using sperm and eggs selected from only the best men and women of the age. Sexual love would wither, but at what benefit! Men would be raised to gods.

(Aldous Huxley wrote *Brave New World* as a rebuke to *Daedalus*. In Huxley's dystopia, factory breeding didn't liberate mankind; it chilled emotion and calcified class divisions.)

For a young American scientist named Hermann Muller, *Daedalus* was a revelation. Muller would be the bridge between the negative eugenics and airy *Daedalus*-style philosophizing of the 1920s and the practical schemes of Robert Graham in the 1980s. Melancholic, attention-seeking, and brilliant, Muller was one of America's first outstanding geneticists. As a junior researcher at the University of Texas in the 1920s, Muller proved that X-rays caused genetic mutations in fruit flies, a discovery that would win him the 1946 Nobel Prize in Medicine. In 1933, the socialist Muller moved to Leningrad to live out his ideals. In the USSR, he drafted a *Daedalus*-inspired eugenic manifesto, *Out of the Night: A Biologist's View of the Future*.

In *Out of the Night*, Muller said that after America's socialist revolution, real eugenics could remake the nation. In a postrevolutionary society, Muller argued, Americans would surely be willing to subordinate their selfish reproductive desires to the common good. A cadre of the best men would be enlisted to be fathers of all mankind. These men would possess the two most valuable human traits: intelligence and "comradeliness"—Muller's catchall term for cooperativeness, good nature, and altruism.

In the current capitalist society, Muller conceded, attempting to breed with the best men would flop. Thanks to distorted American values, women would pick the wrong guys—don't they always?—

producing "a population tomorrow composed of the maximum number of Billy Sundays, Valentinos, Jack Dempseys, Babe Ruths, and even Al Capones." But in a socialist utopia, women would go for Mr. Right (or, rather, Mr. Left). "It would be possible for the majority of the population to become of the innate quality of such men as Lenin, Newton, Leonardo, Pasteur, Beethoven, Omar Khayyam, Pushkin, Sun Yat Sen, Marx." In a few generations, Muller claimed, eugenic sperm banks would enable the number of great men and women to multiply a hundredfold.

Certainly, Muller acknowledged, some people might hesitate at this fundamental change in marriage, but in the end, "how many women, in an enlightened community devoid of superstitious taboos and of sex slavery, would be eager and proud to bear and rear a child of Lenin or Darwin!"

To modern eyes, *Out of the Night* reads almost like a parody in its invocation of Lenin as the ideal sperm donor, its misplaced hope in socialist revolution, its preposterous underestimation of the male ego, and its view of science as a benevolent God, one that can reverse evolution with a flick of its hand. Yet it was when reading *Out of the Night* that I finally began to understand the Nobel sperm bank as something other than a lark. To Muller and his acolyte Robert Graham, this genius factory was nothing less than the most important project of mankind, because it was the only possible salvation of a genetically doomed world. *Out of the Night* was the scaffolding of the Nobel Prize sperm bank, its scientific logic, its animating zeal.

It didn't impress Muller's Soviet sponsors, however. He sent it with a flattering note to Stalin, who loathed it. For that and other reasons, life in the Soviet Union became impossible for Muller, and he fled back to the West lest he be purged. Eventually he settled into a professorship at the University of Indiana.

Despite the disgrace of eugenics by Nazism, the idea of the genius factory continued to entice Muller. With the 1949 discovery of how to freeze and thaw sperm, eugenic sperm banks finally seemed practical. Discarding the socialist idealism of *Out of the Night*, Muller concocted a new justification for the genius factory—a sales pitch for a nuclear age. Muller, whose Nobel-winning experiments had made him the world authority on radiation and mutation, argued that a buildup of atmospheric radiation was altering human DNA at an alarming rate, a rate much faster than evolution could adjust to. Humanity was dooming itself to slow genetic decline as we slowly accumulated bad mutations. The remedy: to freeze the seed of the world's best men in lead-shielded tanks, and use their healthy DNA, instead of our radiation-weakened strands, to breed the next generation.

Muller outlined his scheme for what he called "germinal choice" in a 1961 *Science* magazine article. This proposed a different kind of eugenics than he had preached in *Out of the Night*. No longer was he calling for the state to squirt Lenin seed into women at government filling stations. Now eugenics would be private and voluntary. Families would decide for themselves whether to have mediocre children of their own or glorious ones from the genius bank.

Muller's "germinal repositories" restored some credibility to eugenics. George Bernard Shaw sent a couple hundred dollars in support. Even Aldous Huxley liked the notion, because there was no *Brave New World* problem of state compulsion. Then, in spring 1963, Muller received a manuscript and a note from a man he had never heard of, Robert Graham. Graham asked Muller to write an introduction to his book, *The Future of Man*. Muller read the manuscript and didn't much like it: Graham seemed to care only that sperm donors be brilliant and was indifferent to Muller's belief that they should also be altruistic. Still, Muller suggested some changes to Graham, who took the edits graciously. More important, Muller recognized that Graham was a valuable ally. This was not merely because Graham shared Muller's ideas about impending genetic disaster. It was because Graham shared Muller's ideas *and* was rich.

Muller had been talking about a genius sperm bank for a generation. In Graham, Muller found the man who could make it happen.

Graham invited Muller to California to discuss *The Future of Man* and the potential sperm bank. On June 5, 1963, they met at the Pasadena Sheraton and agreed to establish a sperm bank for "outstanding individuals." Graham went along with Muller's suggestion that the bank seek both "high intelligence and cooperativeness." It must have been a funny encounter: the two aging men—Graham was fifty-six, Muller was seventy-three—planning solemnly to remake the world. Graham fawning over the Nobelist, whom he idolized; Muller bemused by the sycophantic millionaire. They floated names of possible donors: Muller suggested evolutionary biologist Julian Huxley (brother of Aldous), geneticist James Crow, and DNA discoverer James Watson. Graham proposed Muller himself. Graham pledged $1,000 to buy storage tanks and liquid nitrogen and another $300 per year to maintain the bank.

Graham, who had a mania for formality, meticulously recorded an account of their meeting and their decision. Graham wrote: "After more discussion of various aspects of the undertaking, Robert Graham said, 'Let's put it over.' To which Hermann Muller responded: 'Yes.' Thereupon the two shook hands and the project was launched."

The Sheraton pact kicked off a two-year flurry of activity. The idea of a genius sperm bank was slightly outlandish for 1963 America, but not too much so. The United States was enjoying its post-*Sputnik* scientific renaissance, and the egalitarianism of the late 1960s hadn't yet arrived. To scientists, politicians, and journalists, the genius sperm bank sounded prudent, not preposterous. Graham and Muller were taken *very* seriously. They proposed storing the sperm at Caltech, an idea the school contemplated without ridicule. They gathered a distinguished advisory board that included psychologist Raymond Cattell, ecologist Garrett Hardin ("The Tragedy of the Commons"), and Jerome Sherman, the Arkansas professor who had perfected the process of freezing and thawing sperm in 1953. Graham incorporated a nonprofit holding company for the planned bank, the "Foundation for the Advancement of Man." Graham and Muller quarreled about what to name the bank, each trying to compliment the other. Muller proposed the "Robert K. Graham Repository for Germinal Choice." Graham countered with "Hermann J. Muller Repository for Germinal Choice." Graham, the superior flatterer, won, and named it after Muller.

Graham, Muller, and their advisers passionately debated which men were sufficiently outstanding to qualify as donors. Only geniuses? Or geniuses with good politics and big hearts? Graham and Muller contemplated elaborate, government-sponsored panels that would evaluate the worthiness of would-be parents and donors. Muller urged a waiting period: sperm could be released only twenty-five years after the donor's death, so that his accomplishments could be judged worthy by history. The bank was larded with so much bureaucracy and pompous evaluation that it was doomed from the very start.

It was also doomed because Muller and Graham were terribly mismatched. They were working for exactly opposite ends. Graham was an elitist and political conservative. Muller was an egalitarian and socialist—strange traits in a genius sperm banker. Muller had insisted that donors be both smart and cooperative in order to serve his ambition of building a more egalitarian society. But Graham cared not a jot for Muller's interest in cooperativeness. He was going into sperm banking to *prevent* the very socialist utopia that Muller dreamed of. Graham just wanted to breed more Edisons, brilliant men to rule over the bovine masses. Inevitably they quarreled. In 1965, Muller asked Graham to suspend the plans for the bank: better to wait and get it right, Muller warned presciently, than start too soon, be accused of a Hitlerian master-race scheme, and poison the whole project. Graham reluctantly agreed to suspend planning. But two years later, Muller died, and Graham was free of his strictures.

Even so, Graham set aside the genius sperm bank idea for a few more years. In 1971, he officially chartered the Hermann J. Muller Repository for Germinal Choice, but then he did no more. In the early 1970s, Graham handed off day-to-day control of Armorlite to his son Robin. In 1978, he sold the eyeglasses company to 3M for more than $70 million. Graham, who owned or controlled nearly all of it, was rich beyond reason. He took his cash, plowed it into real estate and other investments, and soon found himself with a fortune of $100 million or more.

In 1976, Graham was ready for genius sperm. He had just moved from Pasadena to the ten-acre estate in Escondido. He had plenty of room for the bank and, now that he wasn't running Armorlite, plenty of time. It was the perfect moment, and the perfect place, for him to start. It's no accident that the three most important sperm banks in the world—Graham's Repository for Germinal Choice, the California Cryobank, and the Sperm Bank of California—all began in California in the late 1970s. The state's progressivism and self-improvement ethos made it ideal soil for sperm banks: Customers, libertarian in their sexual and personal behavior, were willing to try anything. And Escondido was the just-right town for Graham's brew of futurism and conservatism. Escondido was located halfway between San Diego, with its defense and biotech industries, and the Central Valley, California's agricultural heartland. San Diego's no-limits futurism was on one side of Graham, the Central Valley's cultural conservatism on the other. With its hills bulldozed into housing developments, its glorious desert valleys irrigated into golf courses, Escondido had the feel of an engineered Eden, a naturally perfect place that man *still* thought he could improve. It was a place where anything seemed possible, as long as it didn't raise property tax rates.

Graham was eager to get started, but he knew nothing about freezing sperm. He was an optometrist. He made inquiries and learned that a young lab technician named Stephen Broder was the man to see. Broder worked at the Tyler Clinic, Los Angeles' leading fertility shop, and he had as much experience banking sperm as anyone did in 1976—which is to say, not much. Graham hired Broder to equip a small lab for him. Broder bought him a few microscopes, some storage vials, and two liquid-nitrogen tanks big enough to hold a few thousand sperm samples. Broder taught Graham how to "process" semen—to measure its potency, dilute it with a preservative solution, and store it in liquid nitrogen. Graham was already seventy years old, but he took to sperm collecting like a boy to baseball cards. He loved fiddling about with the small vials.

The idea of a genius sperm bank made a certain amount of sense, but never as much as Graham dreamed. Graham was making the best of the crude science of his time. If you were hoping to give kids better genes, this was all you could hope to do in the late 1970s. At the time, sperm collection was practically the only widely available fertility treatment that worked. Social science research was beginning to show that intelligence was at least partly heritable. So it was logical that if you were going to have a sperm bank, you might as well select smart men, rather than drag Joe Donors off the street, which is more or less what other banks were doing.

Nothing much was likely to go wrong with Nobel donors, but nor were they the great boon Graham believed they were. Graham thought his donors would supply a massive intelligence boost. In fact, the genetic improvement was probably minuscule. Nobel sperm would give *modest* odds of *slightly* better genes in the *half* share of chromosomes supplied by the father. And even then Graham would be operating on only the nature side of the equation: he had no control over nurture—schools, upbringing, parents. This was a formula for a B-plus student, not the "secular savior" Graham hoped to breed.

Graham puzzled over which men should stock his bank. At first he considered a military sperm bank—only West Point and Naval Academy grads. But eventually he returned to his original idea: the world's smartest men. The best objective measure of *useful* intelligence, Graham thought, was the Nobel Prize—and not just any Nobel Prize but a Nobel Prize in the sciences. He had a narrow imagination about human accomplishment. Graham didn't believe in "multiple intelligences": He believed in *one* intelligence. When he talked about intelligence, all he meant was practical problem-solving ability: Edison, Fulton, Watson, or Crick. (Graham valued, by miraculous coincidence, *exactly* the same kind of analytical talent he himself possessed.) He was blind to the intelligence required for artistic genius, for psychological insight, or for political deftness. Those kinds of intelligence were worthless to Graham, because they couldn't be measured. Scientific ability could be counted in numbers of patents or an IQ score. There was no place for a Picasso or Roosevelt in Graham's Pantheon. A Shakespeare play couldn't light a city at night or fly to the moon. Inventors were the only people who changed the world, and it was their genes that needed saving. (Graham never grappled with a basic contradiction in his own thought: inventions made life more comfortable for the masses, yet Graham believed that comfort was what encouraged the shiftless and stupid to reproduce so rapidly. Thus the better the inventors made the world, the worse the evolutionary crisis.)

Graham always described the Repository as a "genius" sperm bank, yet in some ways he wasn't actually seeking genius. The kind of genius in a Leonardo or an Einstein—incomprehensible, impossible for ordinary minds to follow—was too difficult for Graham. Partly Graham knew, as most scholars of genius have recognized, that it was impossible to manufacture an Einstein. Such transcendent genius arrived uninvited and unexpected, and tended to disappear, too. Einstein left no Einstein-like kids. But Graham also ignored that kind of genius because it didn't match the intelligence he admired: high analytical and technical ability married to hard work. Graham wouldn't have known what to do with an oddball like Einstein. He did know what to do with a dozen engineers.

With Muller dead, Graham felt free to make the bank *he*

wanted to make. Graham scrapped Muller's idea of waiting till a donor was twenty-five years dead before releasing the sperm: the world was going to hell too fast to wait. He also abandoned Muller's altruism requirement for donors, which he had always thought was pointless. So in the late 1970s, Graham fired off flattering letters to all the Nobel Science laureates he could find in California. Their genes were precious, he told them. Could they do a good deed for the world? Would they share their glorious genetic heritage with desperate infertile couples?

When a Nobelist responded to a letter with even the slightest interest, Graham followed up with effusive phone calls to schedule a collection. He took Broder on his collecting expeditions, and both men loved the trips. Broder, still in his twenties, was starstruck when he met the laureates. Graham took pleasure in adding the Nobelists to his lifelong collection of Great Men. (Sometimes the sperm bank seemed a kind of supercharged autograph collection for Graham.) Graham was respectful toward the donors. He always called them "Doctor." He read up on them in advance and asked them polite, informed questions about their work, but not too many, because he thought their time was precious. Even with the Nobelists, however, Graham was unbothered by the inherent awkwardness of asking a man to masturbate in a cup for him.

At first, Graham and Broder collected sperm nearby. They would book a pair of rooms at San Diego's famed Torrey Pines Lodge and fly the donor in. The donor would perform in one room. They would immediately process the sample in the other. ("I don't think we brought 'inspirational literature,' " said Broder, using the industry euphemism for pornography. "They were older fellows and that did not seem appropriate.") Sometimes Graham and Broder had to make peculiar accommodations. Broder had a strenuous arrangement with a donor he describes only as a "world-famous scientist" in Los Angeles. The scientist would call Broder and instruct him to drive by a particular intersection in Century City at a given time. When Broder pulled up, the scientist would open the passenger-side door of Broder's car, drop in a paper bag containing the sample cup, and vanish. Broder would rush it back to his lab and ice it.

In the 1970s sperm collecting was new and mysterious, and Graham and Broder encountered bumps whenever they had to explain what they were doing. In July 1978, for example, Graham and Broder made their first long-distance sperm mission, flying to San Francisco to collect from William Shockley and a second Nobelist. After the Nobelists took care of business—Shockley in a room at the Travelodge—Broder and Graham returned to the Oakland airport toting a white plastic container filled with liquid nitrogen and semen ampules. They had never thought about how they would get onto the plane. X-rays cause mutations, so they couldn't run the sperm samples through the X-ray machine. But they also couldn't conceal the samples in an X-ray-proof lead-lined container, because airline security would reject it. They managed to sneak the container through security and avoid the X-ray machine, but when they got to their plane, the crew turned them away, and no wonder: liquid-nitrogen vapor escaping from the container was swirling out in a ghostly, sinister cloud. Even in 1978, you couldn't carry mysterious, smoking packages on airplanes. The next day, they got on another flight. But before takeoff, the pilot came back to their seats and ordered them off. They begged him, insisted that all they were carrying was a few sperm samples. He relented, and the sperm made it home safely. After that, they shipped by bus and cargo plane.

By 1980, Graham was ready. He had collected sperm from three Nobelists, including Shockley—an "adequate" supply for his first customers. He also had semen from two other revered scientists who weren't Nobelists. According to a former Graham employee, one of those two other scientists was Graham's acquaintance Jonas Salk.

CHAPTER 3
THE SEMEN DETECTIVE

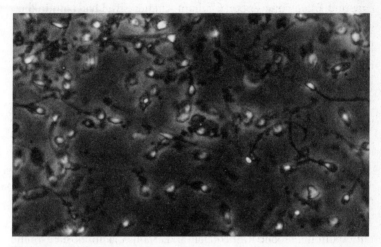

Who's your daddy? *Courtesy of California Cryobank*

I didn't really expect any children, parents, or donors from the Nobel sperm bank to respond to my February 8, 2001, *Slate* article asking them to contact me. I thought I might hear from some fakers and jokers, but that was all. I had read enough about sperm banks to know that most parents never told their kids that they were donor babies, so why would they spill the secret to a reporter? Moreover, the Repository had had very few children and donors: the bank had produced only about two hundred children and employed only a few dozen donors. The chances of any of them—let alone the few who'd be willing to talk—seeing a *Slate* article seemed vanishingly small.

As I waited anxiously to hear from them, I wondered how the reality of the Nobel sperm bank would compare to my own fantasies about it. Infected by Robert Graham's grandiosity, I had conjured up creepy Boys-from-Brazil visions about the kids: What if they really *were* a regiment of amazing but scary children, pawns in a futuristic plot to alter humanity? What if the Nobel sperm bank parents were raising their enhanced babies in geodesic domes off the grid in Montana—Mensa moms nourishing their genius toddlers with a diet of brain-stimulating vitamins, Shakespeare, and calculus, biding their time till the genetic elite was old enough to seize control of the nation? It was possible, wasn't it?

Less than twenty-four hours after my article was published, I received an e-mail from a man I'll call Edward Burnham. He wrote that he was one of Robert Graham's donors, and he wanted to tell me about it. I replied immediately and arranged a phone date for the next day. When I called, Edward immediately confirmed what I already knew from having done an Internet search on him: he wasn't a Nobel laureate. I considered the possibility that he might be a fraud, but he was prepared for suspicion. He supplied me with names of bank employees, dates of his donations, his Repository medical records, his donor catalog entry, and even a leftover plastic vial, a souvenir from his last donation a decade ago.

Edward wasn't a scientist, but he was the other type of man that Graham revered: a self-made businessman. He was in his late forties, and he had built three companies from the ground up and sold them. In 1985, Edward told me, Graham had seen him speak in San Diego. Graham had immediately set his cap at the young millionaire and invited him to donate to his sperm bank. Edward, assuming that it was still the "Nobel" sperm bank, wondered why Graham was pursuing him. Graham told Edward that he was no longer recruiting Nobel laureates. Instead he was now seeking donors more like

Edward—Renaissance men who were smart, but also young, athletic, and handsome. Then in his early thirties, Edward reminded Graham of himself. Edward, too, was a rational entrepreneur. Edward's companies, like Graham's Armorlite, turned theoretical technology into real profit. When I met Edward on his Arizona ranch months later, I understood the other reasons Graham had lusted after him as a donor: Edward looked the part of a superman, with a powerful chin, wonderful cheeks, piercing brown eyes, a mass of curly black hair, and the build of a linebacker. He also raced motorcycles and played soccer. (And he was a mensch. He told jokes, he made fun of himself, he asked questions.)

Edward refused Graham's first invitation, but the sperm banker wouldn't stop bothering him. Graham flattered Edward so much that "I started to feel like the dog at the dog breeders' meeting," Edward said. Edward had a girlfriend and he had pets, and they were obligation enough. He hated his own dad and had run away from home. Edward didn't want children, didn't much like them, and saw no particular reason to produce more of them for anyone else. But Graham would not relent. Edward could make some worthy women happy, Graham insisted, and he would bear no responsibility for the kids. Edward's then girlfriend—hoping to get married and have children with him—lobbied him, too. She thought that once Edward donated sperm, his resistance to fathering his own kids would crumble. Eventually, Edward caved. He scoffed at Graham's grand eugenic dreams—"I thought it was pissing in the ocean"—but he figured he would make a few women's lives better, and his girlfriend and Graham would stop pestering him. To his surprise, Edward discovered he didn't mind donating, and he gave for more than five years. (The girlfriend's dream was dashed, though: he *still* didn't want kids, and he *still* didn't want to marry her. When Edward, his new girlfriend, and I went out to dinner, she told me she had resigned herself to not having kids with Edward but added wistfully, "I *know* we would have had beautiful intelligent babies, genius sperm bank babies.")

Graham told Edward that his sperm had produced several children, but Edward never asked exactly how many. He didn't want to know. He told me he never thought about the kids. Did he care what became of them? I asked. Not really, because they were not his.

A couple of days after our first conversation, I published an interview with Edward in *Slate*. Edward wouldn't let me publish his donor color code, for fear that parents or children would try to hunt him down, so I gave him the pseudonym "Donor Entrepreneur." I feared that Edward's story—highlighting as it did his doubts about the Repository's goals and his indifference to his biological kids—might discourage anyone else from contacting me.

Instead, Edward's story shook the trees. Within days, I had heard from half a dozen other donors. Several of them had seen Edward's comments and wanted to rebut him. Unlike him, most were true believers. They did think the Repository made a difference. It was not ocean pissing.

The notable fact about these donors: none of them was a Nobel laureate, either. Why did the "Nobel Prize" sperm bank not seem to have *any* Nobel Prize donors? I wouldn't learn the answer till later.

These donors were a long, long way from the Nobel Prize, in fact. They were a motley lot, not exactly unimpressive but certainly not the great minds of the age. A couple were university professors. There was a former math prodigy—a cheerily self-proclaimed "failure"—who had quit academia to become a sculptor. There was a graduate student who had aced his SAT. There was another former prodigy, who had a gigantic IQ but whose job was doing extremely low-level work for intelligence agencies. And then there was a political activist of a particularly loathsome stripe. I recognized his name from news stories about his repellent ideas. Graham had recruited him at a political conference. (The thought that there could be more kids with his genes growing up across America filled me with dread. I was relieved when he told me that the sperm bank had rejected him for unspecified medical reasons.) All the successful donors said they had been flattered when the Repository had asked them to participate and had contributed eagerly. (One of them had *volunteered* his services to Graham.) Unlike Edward, these donors did want to hear about their offspring, maybe even meet them.

There was one Nobel connection among the donors who contacted me, but it wasn't what I expected. I received a note from "Michael," who said he would be glad to tell me about being a donor. Michael's last name sounded familiar, and when I asked him about it, he said that, yes, he was the son of Nobel prize–winning scientist _____ _____.

I was intrigued. How had a Nobelist's son become a Nobel sperm bank donor? Michael was very cagey—living in his father's shadow seemed to trouble him—but after much back-and-forth, he agreed to let me visit him in Dallas. He picked me up at the airport driving a wreck of a car. Michael was about fifty years old. He was tall and alarmingly gaunt: when he folded himself into the driver's seat, he looked like a praying mantis. He spoke in a high-pitched, hesitating voice that immediately grated on me.

We drove back to his condo. His wife wasn't home. He directed me to sit down in the living room. Except for small islands around the couch and TV, every inch of floor space was covered with flowering plants. Their perfume was almost overpowering.

Michael, I learned, was a man of little renown. It was not clear whether he had a job. He taught piano occasionally. But it was obvious within minutes of meeting him that Michael's true calling was donating sperm. He referred to sperm donation, unironically, as "work." As in "I continued to *work* for that fertility doctor, but I looked around for more *work.*" He was the only person I have ever heard of—outside the porn industry—who thought of masturbation as labor.

Michael didn't mean that sperm donation was "work" in the sense that he did it to earn a living. All the banks except the Repository had paid him a little, but he would have done it for free. No, he called it "work" because it was the most productive activity in his life.

Michael said he'd gotten hooked on sperm donation in the mid-1970s when he'd read an article in *Playboy:* "You can be a sperm donor." He had just finished his music degree and was living in the Northwest. He soon found plenty of "work." "I started calling around to obstetricians and gynecologists until I found one who wanted me." He had "worked" for the first doctor, then found others and started supplying them, too. (In those days, before large sperm banks were so popular, many doctors collected and distributed their own sperm.) As we talked, he ticked off his employers on his fingers. "Oh, there were probably half a dozen doctors I worked for, plus two or three sperm banks." All in all, he had spent fifteen years masturbating. It had, he admitted, been exhausting.

Michael said he had volunteered himself to Robert Graham in the mid-1980s. Graham, though he had stopped recruiting actual laureates, was thrilled to add Michael's second-generation Nobel genes to his bank. The Repository catalog hinted coyly at his Nobel heritage, describing an accomplished musician with an "outstanding history of achievement . . . in his family." (Michael acknowledged what I had suspected: his father had been one of Graham's three original Nobel donors, though he hadn't stuck with it. Because the dad had quit, Graham might have been more eager to bank the son's seed.)

Michael and his wife had no children of their own, but Michael's eagerness to reproduce had not faded with age. The only reason he had stopped donating was that he was now so old that sperm banks wouldn't accept him anymore. He tried to work around the age restriction. He recently learned that one sperm bank he had donated to was merely *storing* his sperm for the future, not distributing it to clients, so he pestered the bank to return the stored samples to him. He wanted to give away the samples himself. He placed ads in a local newspaper volunteering his sperm to lesbian couples and single women. He was hoping to find a woman who would let him stay in touch with the child. Not that he intended to financially support the kid or be a father—he just wanted to check in when it was convenient. When I wondered how many kids he had fathered, Michael stopped to pause and calculate. He guessed fifty, including fifteen through the Nobel sperm bank.

I asked him why he had spent the best years of his life donating sperm. Michael lit up. "When I heard about being a sperm donor, I thought, this is *great*! I am helping women. I am helping the human race because I have good genes. And I am passing on my genes."

He leaned in, his voice urgent, his skeletal fingers pointing at me. "I have studied evolutionary biology, and *this* is what evolution is all about. Winning is passing on your genes, and losing is failing to do so. There are lots of games that men have made up, games where you win by scoring runs." He paused, as if to emphasize the pointlessness of such games. "But the *main* game of the universe, the *only* game that matters, is the game of evolution, and you win by passing on genes. And I wanted to *win*!" He spoke this last sentence with a smug grin. It was just about the creepiest thing I have ever heard anyone say.

Sitting in his depressing condo, I looked at Michael and thought, You *are genetic victory*?

Michael, I realized glumly, was the living test of Robert Graham's theory. He was the son of one of Graham's original Nobel donors: in other words, exactly the kind of person Graham aspired to create. Michael was the finished product of Graham's logic—blessed with allegedly magnificent Nobel genes. Yet all I saw in him was the fickleness of DNA: Here was a Nobel Prize baby, and he was no prize at all.

While I was talking to donors, I began to hear from parents with children by the Repository as well. The first to e-mail me were mothers who had read about Edward. I wrote about these mothers in *Slate*, and other parents saw those stories and contacted me, which led to more articles. Each story attracted another few parents and kids, until eventually I was in touch with two dozen families. The Internet did exactly what my editors and I had hoped: it allowed readers to collaborate with me to discover the lost history of the Repository.

The first Repository parent I actually met was "Lorraine O'Brien." A couple weeks after my first article, I traveled to see her. I had found Lorraine, rather than her finding me. I had seen her name in an old newspaper article about the Repository. Lorraine was taken aback that I had located her—the newspaper reporter had been supposed to keep Lorraine anonymous, and Lorraine had never seen the finished article with her real name in it—but she agreed to see me anyway.

Lorraine was a neurologist, and I dropped by her exceptionally busy practice one afternoon in February. Lorraine was a brunette in her early forties. She was funny, fast-talking, conservative, beautiful, and ruthless. She was full of advice, most of it smart, all of it absolutely certain. Lorraine said she had three kids, a ten-year-old boy and six-year-old twins, one boy and one girl. The Nobel sperm bank's "Donor Fuchsia" had been the father of all of them.

Many of the moms I was hearing from were doctors, nurses, or psychologists. As I talked to Lorraine, I understood why this was so. Lorraine said that she saw the miserable consequences of poor health every day at work. So when she went shopping for sperm banks in 1990, she wanted to ensure her donor was a Grade A specimen. The Repository told her far more about its donors' health and accomplishments than any other bank would. The Repository let her interview its manager and peruse the catalog. Eventually, she concluded it was the only bank rigorous enough to father her children.

But health wasn't the only reason Lorraine had chosen the Nobel sperm bank. She was also an unabashed elitist. "When you are growing fruits and vegetables, you don't pick the bad ones and try to grow them. You pick the best. Same with kids."

She went to the Repository because she wanted superkids, and as far as she was concerned, she got them. The very first thing she told me about her kids was "They're wonderful children. They are like royal children." She thrust pictures of them at me. Other parents, she declared, loved her children. Friends volunteered to babysit them: "Do you know what kind of kids you have to have for people to *volunteer* to babysit them?"

Then she announced, "I'm a great mother," as if daring me to contradict her. Her ex-husband was no father at all to the children, she said. "I am the exclusive provider of food, money, shelter, emotional support. . . . They do not have a bond with him." She more than compensated for him, she said, by being such a relentless parent. "When my kids were babies, some other moms and I—all of us obsessive-compulsive women—formed a group called 'better baby salon' to raise morally intelligent children. . . . I read everything about raising children, good and bad, because you have to take care of that yourself."

Her kids had become paragons of achievement. Here are some of my notes from the conversation: "99th percentile . . . best school in the county . . . student of the month . . . little angel . . . their Valentines were the best in the class . . . her coach said he had never seen a faster runner . . . unbelievable . . . never had in my entire teaching career a child who is as emotionally balanced as this child . . . enrichment classes."

The Nobel sperm bank was intended as a scientific experiment to prove how nature trumped nurture. But Lorraine was evidence why the bank could never show that. Her kids might outperform regular kids, but that proved nothing about heredity versus environment. Graham's customers had not been randomly plucked off the streets of San Diego. They had *chosen* the Nobel sperm bank, and they were the kind of parents you would expect to pick a Nobel sperm bank. Lorraine had gone there because she cared passionately that her kids be standouts. Even if Lorraine had gone to Tony's

Discount Sperm Warehouse, her children would have been achievers: she wouldn't have let them be any other way.

I don't mean this negatively; hyperinvolved parents are often the best kind. I hope I can do as much for my kids' lives as Lorraine did for hers. But I didn't want to mistake her children's accomplishments for genius. I am skeptical of the high-achieving child, perhaps because I was once one myself. I tested outrageously well. I aced my SATs at age eleven. My teachers were always predicting a glorious career in mathematics, medicine, or philosophy. But what looked like genius was simply good parenting. I was a pliant, reasonably intelligent, eager-to-please child with bright, attentive parents. Of course I did well in school. By the time I hit college, when I met the real geniuses, the people with incomprehensibly dazzling minds, I recognized I wasn't one of them or anything like them. So I knew just how little being a ten-year-old prodigy meant—and how cautious I should be about ascribing any child's accomplishments to the Repository's superior sperm.

As Lorraine boasted about her kids, she also kept insisting that they were "normal"—or, as she put it, "NORMAL!" They were "*not* nerds," she said emphatically. Lorraine's insistence on normality at first befuddled me. But eventually I traced it to the deeply democratic habits of Americans, so deep they still register in a woman as elitist as Lorraine. Graham's notion of breeding only for intelligence disturbed even her. If all men are created equal, then manufacturing an extraordinary child seems almost anti-American. America admires its Thomas Edisons and Bill Gateses when they grow up, but Gates-like kids are ridiculed. The *genius* child is considered a nerd or a freak. Instead, we cherish all-arounders. There's no glory in being a math prodigy, but a math prodigy who can play basketball, that's cool. That was why Lorraine kept battering me with that word "normal" and why Graham's Nobel effort had been quixotic: he was trying to sell a product—pure intelligence—that most Americans didn't really want. Even Lorraine, a mother of immeasurable ambition, didn't want me to perceive her children as too intelligent. It was when I talked to Lorraine that I started to understand why Graham had to recruit non-Nobelists like Edward Burnham as donors: parents really *didn't* value the Nobel brain above all else.

For a while, Lorraine and I avoided talking about one obvious subject: Donor Fuchsia. I sensed some anxiety in her about him, some tension in her otherwise assured manner. Gradually, she told me little bits about him. He was not a Nobelist, but he was an Olympic gold medalist. He had also written a book. (That was why she had chosen him, because he was well rounded, "not weird or nerdy.") Finally, forty-five minutes into our conversation, she blurted out, "I have seen pictures of him."

Then she said, "Dora Vaux [the Repository's manager in the early 1990s] told me his real name."

I stuttered that revealing his name to a mother was surely against the rules. Lorraine, who would probably break the law of gravity if it displeased her, agreed but said that was irrelevant because she had really wanted to know it.

Then Lorraine told me his name and how he had won his gold medal. I felt guilty even *hearing* the name: Lorraine shouldn't have possessed this secret, and neither should I. I steered the conversation back to her kids, but a few minutes later, she mentioned Donor Fuchsia again. "I have a *huge* file on him, you know. I have not used it for anything in particular. I am just curious about what he is doing now."

A few minutes later, she said, "Supposedly, the guy is just my age. Dora told me that his girlfriend got really sick, with leukemia, and he stuck by her. And he helped Dora move. She says he is the most incredible person.

"And he's *not* married. He never got married."

Again, a little later, "I thought about calling him, but I don't know what his feelings about this are. I thought we might meet serendipitously and fall madly in love, and he would become the father of his own children. That would be a movie and a half!"

Then, as she was saying good-bye, "I thought he and I might meet someday. Wouldn't that be a story and a half?"

When I heard this, I thought I finally understood why Lorraine had agreed to talk to me: she hoped I would find Fuchsia for her. I doubted that this was a conscious plan, but it explained why she kept telling me more than she should have about him. And it explained why she had kept hinting about them meeting. She realized that she couldn't track him down herself. It would be too awkward. But it wouldn't be awkward for me. I had the reporter's excuse for calling him. If he didn't want to talk to me, no one's feelings would be hurt, no one's ego bruised.

So when I got back to Washington, I set to work locating Fuchsia. It's harder than you might think to find an Olympic gold medalist, but eventually I turned up a cell phone number. I called him. I told him I had heard he had been a donor to the Nobel sperm bank. I reassured him that I wouldn't reveal his name. (I did *not* tell him "A woman who has three of your children knows your identity. She might like to meet you and fall in love and have you become a real father to your children. And wouldn't that be a story and a half?")

He listened quietly. He didn't deny that he was Fuchsia, but he asked to continue the conversation by e-mail. I sent him an e-mail, and he replied the next day that after serious consideration he had decided not to talk to me, or to "participate in [my] endeavor."

At the time, I didn't tell Lorraine about his refusal to talk. What would it have done except spoil her dream? When I saw her again two years later, I did tell her. But she had a new boyfriend then and seemed unbothered that Fuchsia didn't want to meet his children or her.

The failure with Fuchsia marked the beginning of my new and entirely unexpected career: sperm private investigator. My encounter with Lorraine made me realize that the real reason she and other mothers would talk to me was *not* to validate Robert Graham's eugenic sperm bank or even to brag about their children. Rather, they wanted my help. There was an emotional void in their lives. They dreamed of finding donors and half siblings for their kids. Until I came along, they'd had nowhere to look. The bank was closed; there were no public records. Thanks to my project, I was learning the identities of donors, children, and parents, information that was supposed to stay private forever. They realized before I did that I might be able to introduce people who hadn't been connected except when sperm met egg five or ten or twenty years ago.

When I had started trolling for the kids, mothers, and donors, I had assumed the traditional, self-serving role of the journalist: I was going to use *them* to tell the real history of the Nobel sperm bank. But instead, I discovered that my sources wanted to use *me*. They had reversed me. They put me to work for them. Mothers asked me to hunt for their donors. Kids asked me to hunt for their fathers and half siblings. Donors asked me to hunt for their kids. It wasn't journalism anymore. I became—through unique access to information, through moral obligation, and through my own curiosity—the Semen Detective.

CHAPTER 4
DONOR CORAL

I returned from my California visit with Lorraine to find a cryptic message on my voice mail. It was a boy with a slow, gravelly voice. He said he needed to talk to me, but he didn't leave a number. He didn't leave a name, either—or rather he left two names, first "John," then "Tom." They both sounded fake. Then I saw this in my e-mail inbox:

> I dont want my real name revealed through this just call me Goldie its a nickname i have kept for a while among friends. about one week ago my mother informed me that my half-sister and i were both one of the 230 who were born through artificial insemination and the Repository for Germinal Choice. I am searching for the papers that we have showing the "color" of the donor, when i find them i will email you again, until that time the only information that i have about my father is that he was one of the few nobel prize winners who gave sperm, if you could give me a list of possible nobel prize winners who could have been the ones who donated, i

would be much appreciative. Then when i find the papers i
will tell you the color of my sister and my fathers and you
can do whatever you do with them. Thank you for all your
assistance as i know you will help in any way possible.

—Goldie

The e-mail fascinated me. "Goldie" was the first Repository kid
who had reached me on his (or her?) own. Some of the Repository
moms had let me e-mail with their children, but those exchanges had
been dreary. The kids had been all preteens: they'd had nothing to say,
and their moms had clearly been lurking, making sure they didn't
write anything interesting. But Goldie was a kid who sounded like a
kid, all bad grammar and enthusiasm.

I figured "Goldie" and "Tom" and "John" were all the same person. The voice mail and the e-mail shared a scattered, frantic eagerness. I e-mailed a reply to Goldie with lots of questions: Are you a boy
or a girl? How old are you? Where do you live? Are you the kid who
left me a voice mail? Because he had asked for the names of Nobelists,
I sent him a Web page that listed laureates, but I also warned him
about what Edward Burnham had told me, that Graham had stopped
recruiting Nobelists very early. I added, gently, that perhaps he should
lower his hopes, because as far as I could tell, there had not been any
Nobel Prize donors after the initial batch of three. I asked Goldie if
we could talk on the phone.

He quickly e-mailed a reply. He said that, yes, he had left the
cryptic voice mail, and his real name was Tom Legare. He said he
lived in a suburb of Kansas City. He had just learned that he was a
child of the Nobel sperm bank, and he had come across my articles
in a Web search. He said he suspected that Jonas Salk could be his
dad. Salk wasn't a Nobelist, Tom realized, but he had seen Salk's
name on the Repository stationery in his mom's files.

He gave his phone number. I called it and asked for his mom,

Mary. I wanted her permission before I spoke to Tom, who was still a minor. Mary said she wouldn't normally talk about the Repository with a stranger, but she figured that it would be worth telling the story if I could help Tom find his dad. I prodded her with questions. I asked if she had wanted a superkid. I wondered how Graham's exotic sperm had made its way to an obscure midwestern suburb. Mary wasn't like the other Repository moms I had talked to. She didn't live in some swank Los Angeles home; she had spent her entire life in suburban Missouri. She wasn't a doctor or nurse or psychologist. She had started in the working class, and, only now, thanks to decades of hard work, had she clawed her family into the middle class.

Mary was in her early forties. She said she had gotten a bad start. She had married at nineteen. Her husband, Alvin, was a Vietnam vet whose work history had been spotty. Finally, he had found work that suited him, in sales. But he was on the road, away from the family for weeks at a time. But Mary was a striver. She'd begun as a secretary at the local insurance company. The company had paid for her associate's degree, then a bachelor's in computer science. She'd inched her way up the career ladder. Now she did tech support for the firm and taught computer classes part-time at the local community college. She owned a nice house. She was doing well.

Mary had stumbled onto the Repository, she said. After seven years of marriage, she couldn't get pregnant. Alvin told her he had been kicked in the testicles in Vietnam—maybe that was the problem. It was 1984: there was no way to fix him, certainly no way they could afford. His doctor suggested the Repository. Mary had never heard of it, but she immediately decided a Nobel sperm bank was a tremendous idea. She was a self-improver: What bigger kick up the ladder could there be for her kids than exchanging her husband's mediocre genes for a Nobel Prize winner's? She was pretty sure the bank had told her that her donor was a Nobel laureate, but even if he wasn't a Nobelist, he was surely better than Alvin. The $500 deposit for the liquid-nitrogen tank had been a ton of money for her back then, but she hadn't hesitated. She'd conceived Tom almost immediately and given birth to him in 1985. When she'd returned for a second child a couple years later, the bank had run out of sperm from Tom's donor, so Mary had chosen a different donor, and that had produced her daughter, Jessica. Mary told me she had raised the kids by herself. She and Alvin hadn't divorced, but his chief contribution to parenting was to watch TV with the kids once in a while. She was a bullying mother, she proclaimed cheerfully. Jessica had been enrolled in dance classes as soon as she could walk. She nagged Tom and Jessica because they were not as driven as she was. If Tom brought home a C, God forbid, she wondered why it wasn't a B. If it became a B, she hectored, why not an A?

As they grew up, Mary had guarded their secret as if it were a treasure. She loved romance novels. She had even written one when Tom was little and still dreamed of finding a publisher for it. From romances, perhaps, she had learned the value of a secret revealed at just the right time. *Wait—that handsome roughneck with a heart of gold who saved that little girl from drowning in quicksand, that's really multimillionaire businessman Lance Stone?!?* She had waited patiently to spring the news on her kids: You're not who you think you are! You are a *Nobel Baby*! She planned to strike a well-timed blow that would change their lives forever.

The moment had just come for Tom, she said, because he had been taunting her with talk of pro wrestling school. It was time Tom knew he was not doomed to become a wastrel. He needed to understand that his genes said—perhaps even *dictated*—that he could become great.

Mary gave me permission to talk to Tom, so I called him the next night. For a while I just listened to him riff about his life. I had to strain to hear at first—his voice was so mumbly—but I soon realized he was funny and self-mocking. I especially liked how easily he rose to indignation at his mom and dad (antiparental indignation being the existential state of the fifteen-year-old boy). *Wrestling school*, he sneered in his mom's direction, that was a joke. Didn't she realize he was planning to go to college *and* wrestling school at the same time? Didn't she realize that what he really cared about was rap, not wrestling? He said he had a band called Infernal. He wrote all the lyrics. He also kept saying "ICP" this and "ICP" that, and I had no idea what he was talking about until I remembered—some neurons firing in the deep reptilian corner of my brain that still read *Rolling Stone*—that there was a band called Insane Clown Posse—a hard-core white rap group. Its music is a dog whistle tuned to the testosterone-addled frequency of the overdramatic teenage white boy. Tom just loved Insane Clown Posse. Tom also said he played a ton of video games, particularly first-person shooters like Halo. He said he was a Wiccan.

Tom told me he had gotten into trouble at school because he had written a song about suicide. The song was antisuicide—a friend of his had been suicidal and the song had been a wake-up call—but someone had narced him to the principal because Tom's lyrics mentioned Columbine. The principal had called the cops. The cops had wanted to know if Tom was planning to shoot up the school. They'd said they were going to arrest him, but instead they'd just sent a file on him to state police headquarters and given him a warning. Now some of the teachers were looking at him funny and had said he better not try anything in class, and it was all because some idiot couldn't understand that these were just song lyrics and that the lyrics said the opposite of what the principal thought they did.

Grudgingly, Tom talked about classes. He said he managed a 3.6 GPA but didn't try. He was a sophomore, but he was already taking courses at the community college. By the time he graduated high school, he would be most of the way to a junior college degree.

Finally I asked him about the sperm bank. He had known about it for a couple weeks by then, and he was still perplexed, toggling be-tween amazement and annoyance. He kept saying, "It's so science fiction"—which was praise coming from Tom. But in the next breath he might denounce it as a "Nazi" project because, as he had learned, all the donors had been white and lesbians hadn't been allowed to apply.

Tom told me how delighted he was to be free of his dad, Alvin—"not the most attractive character," he said dryly—but he still didn't know what to think of his new, nameless, long-gone biological dad.

Mostly he raged at his mom. He hated that she had kept the secret so long. He hated that she had misinformed him that his donor was definitely a Nobelist. He hated that she couldn't find any of her Repository records. He hated it even more that when Tom himself had managed to find the records, she couldn't remember which guy was his dad.

Tom said he felt like he was worse off than he was before he knew the truth. "At first when she said it was a Nobel Prize winner, I figured there weren't that many, and I would be able to find him. But if it's not a Nobel Prize winner, if it could be *anyone*, then I will never find him. And so now I have no father. My father—my mom's husband—isn't my father. My *real* father—the donor—isn't my father because all he did was donate sperm, which is not enough to make him a father. So nobody is my father."

As we ended the phone call, I promised Tom I would do what I could to help but that I wouldn't be able to discover anything without the donor's color ID. I had realized by then that the color code was the key to matching kids with donors. Tom and Mary put their heads together again and re-examined the donor catalog. Looking a second time, Mary suddenly remembered: maybe it had been Donor Green. He was a "Professor of a hard science at a major university and already one of the most eminent men in his field." His IQ had been measured over 200 as a child. He had "extraordinary powers of concentration," "seldom loses his temper," and "enjoys playing with children, folk dancing, and linguistics." Mary suspected that Jessica's donor had been Turquoise ("a top science professor at a major university, head of a large research lab . . . a professional musician"). She and Tom asked me how they could confirm her suspicions.

I suggested that Mary contact Marta Graham, Robert Graham's widow, who might have kept the records when the bank closed in 1999. Perhaps Marta could confirm which donor Mary had used. I found a phone number for Marta Graham, and Mary called her. Mary explained that she believed that telling her son something about his donor might inspire him to do better in school. But Marta Graham said she couldn't help, because she didn't have the records. Mary eventually located the person who did have the records, a San Diego woman named Hazel. Mary wrote Hazel a letter asking for the donor code names. Finally, in summer 2001, six months after Tom and I first talked, Hazel mailed an answer. Mary had misremembered. Jessica's donor hadn't been Turquoise; it had been Fuchsia, the Olympic gold medalist, who was also the father of Lorraine's three kids. Jessica didn't know the secret yet, so Mary just filed Fuchsia's code name away.

The second surprise from Hazel: Tom's father was not Donor Green. It was Donor Coral. Mary and Tom were thrilled to have the name, or at least the code name. According to their catalog, Coral was a gem: "A professional man of very high standing in his science, has had a book published." His IQ had been tested at 160 at age nine. His hobbies were "writing, family, reading, chess and piano proficiently." He was described as good-looking, happy, easygoing, extroverted, and wonderful with kids. His health was superb. He excelled at all sports. He came from a large family of "high achievers." And he was young, born in the 1950s. That birthday ruled out the elderly Jonas Salk, but Tom didn't care. Coral sounded perfect—even better than Salk.

Now that he had learned something about his "real dad" (as Tom was now calling Coral), Tom told his dad, Alvin, that he knew the secret. He wanted to see how Alvin responded. But the conversation was an anticlimax. His dad responded with his usual sullen indifference. They didn't talk about it again.

Giddy on Coral, Tom also broke his promise and revealed the secret to his sister. Jessica was less shaken than Tom had been to discover she was a genius sperm bank product. She was also less surprised. She was generally hard to faze. Her relationship with Alvin had soured, too, so she was glad to know he wasn't a blood relative. But she didn't feel the same compulsion as Tom to search for her donor dad. She was only a little curious about Donor Fuchsia. She wondered how old he was and what sport he played, but that was it. The void didn't bug her. Mostly she was happy because learning the secret brought her close to Tom again. Oddly, discovering that they were *less* related by blood—they shared only one parent, not two—made them feel more like brother and sister. They'd barely talked in years, but suddenly they were friends again.

Tom called me with the Coral news. He told me his image of what Coral might be like: "I think he's a writer. He probably lives in California. He is probably married and has kids of his own."

He asked me to help him keep looking for Coral. "I just want to see what he looks like, to find out what he did with his life, and maybe to talk to him, just once. I don't know whether I want to tell him, 'How dare you! How could you have done that?' or tell him, 'Good job, thanks.' I won't know till I see him."

Tom also wondered if I could find any Coral siblings. He would love to have sisters and brothers. He asked if I had been contacted by any other Coral kids.

In fact, I told Tom, I already knew his brother. He was fourteen. He lived in Cambridge, Massachusetts. His name was Alton Grant.

I had heard from Samantha Grant, Alton Grant's mother, six months earlier, right about the time I had first talked to Tom and Mary. Samantha was one of the first mothers who had contacted me. It happened in a roundabout way. First I had received a mysterious e-mail from someone claiming that she knew a Repository family. A few days later, a woman left a vague message on my voice mail, but with no name or call-back number. Soon after that, Samantha sent me an e-mail:

> You know me in several guises already that you may think
> are independent. Confidentiality is a serious issue and so I
> have approached you cautiously, from different directions,
> to test the waters. I am ready to talk with you, and perhaps
> meet with you.

I e-mailed her back. She said her niece had sent the first e-mail, feeling me out on Samantha's behalf. Samantha and I arranged a time to speak. I liked her immediately. She was a chief engineer at a Boston high-tech company, and she was smart as hell. Samantha had grown up in a place—the rural Midwest—where smart girls were viewed with suspicion. Samantha had accepted the cost of this— being viewed as odd by narrow-minded people—rather than rein in her brain. She reminded me a little of my mother, who's an English professor. My mom's a generation older than Samantha, but they're both endlessly curious women who are intellectual without being the least bit snobbish. Samantha had the rare gift of being able to explain difficult ideas clearly and the even rarer gift of believing that those ideas should be taken seriously, argued, and celebrated. Samantha knew that I knew nothing about her field—a particularly challenging branch of engineering—but that never stopped her from describing her work to me and never stopped me from enjoying it when she did. She told a story well and she was very funny. She started with a belief in the goodness of all souls, but when someone exposed himself as an ass or a monster, her tongue was acid.

Samantha was fiercely protective of both her own privacy and her son's. Our first conversation, before she would reveal a jot of information, was a negotiation about privacy.

Finally, she told me her story. It started in the mid-1980s. "I wanted a child, but my then-husband had had a vasectomy." At first, she asked a friend of a friend if he'd be willing to donate sperm to her. He lived overseas, had kids of his own, and would stay out of Samantha's life. But she scrapped that plan, because the "known" donor proved too emotionally entangling. Then Samantha read an article in the *Los Angeles Times* about the Repository for Germinal Choice. Her inner nerd loved the idea of a genius sperm bank. She was smart, she'd always been smart, and she wanted a smart kid. She thought that she understood how to raise a smart child, but that she would have no idea how to raise a jock or a beauty queen. Samantha was the only Repository mom I talked to who was not afraid of genius, perhaps because she had so much of it herself. "I can't understand why anyone would think it is bad to want to have a bright child."

She ordered the Repository catalog in early 1985. Lots of donors tempted her, but it came down to a choice between Coral and Light Green. She couldn't make up her mind. She was living in California then, so she drove down to the Repository office in Escondido. The Repository office manager, Julianna McKillop, sold her on Coral. Julianna told Samantha that Donor Coral was happy. She said he was highly accomplished in mathematics. The engineer in Samantha liked that. Julianna said Coral's sister was a world-class pianist. Samantha, who'd almost become a professional violinist, was delighted. Julianna said Coral's parents were wonderful people. She said he loved children and had three children of his own. She said he was a head turner. Then Julianna pulled out a picture of Coral and showed it to Samantha. He was bright-eyed, floppy-haired, and cute. "That sealed it. Everything I heard, I liked."

She ordered a liquid-nitrogen tank full of Coral. As she drove home, the tank riding shotgun in her car: "I was thinking, 'My kid is sitting next to me!' "

In 1986, Samantha's son, Alton, was born—a year later and 1,500 miles west of his half brother Tom. Alton, she said, had grown up a very happy child, and a gifted one. Prodded by me, Samantha recited a litany of his achievements. He was a first-class pianist. He had had a piece of sculpture in a children's art show at Harvard. He studied dance. He was interested in marine biology. Samantha said she offered her son every intellectual opportunity she could: "I do expose him to great minds whenever I can, and great books and music. These inspire him to seek deep levels in whatever he does." But Samantha said that he drove himself, and she had to restrain him from doing too much.

Alton wasn't close to his father, who was recently divorced from Samantha. Around the time of the split, Samantha had told Alton that Coral was his real father. He had been unperturbed, she said. He had told her he was relieved and that he had always known his father wasn't his father, even if he hadn't *known* it. But Alton had not asked her any questions about Coral and expressed no interest in finding and meeting him.

It didn't surprise me that Samantha was a divorcee. Almost all the parents I heard from were mothers who had divorced or were planning to divorce. This made sense: Married couples would be much less likely to share the secret that their children were the result of sperm donation, because the husbands usually pretended to be biological fathers. With the husbands out of the picture, divorced mothers were much more willing to share the Repository secret with their children, and even a stranger like me. There was a second reason why newly single mothers tended to seek me out. The divorces had shrunk their families. They hoped that I could help them find new relatives for their kids, either half siblings or donor fathers. The intact families, by contrast, weren't searching for anything.

Samantha and I struck up an energetic correspondence. We e-mailed a lot, mostly about Repository business or my latest story, but we also chitchatted about her work or Alton or my daughter. So as soon as Tom told me he was a Coral boy, I e-mailed Samantha the news. She answered instantly, "Wow, is there any chance of connecting with them?"

I gave Samantha Mary's e-mail address. The two moms corresponded briefly, each bragging about her son (his college courses, his hobbies) and asking lots of questions about the other boy. They agreed to let Tom and Alton—the brothers—talk to each other.

I was curious—and anxious—about what would happen. As far as I could tell from reading professional literature and newspaper articles, this was only the second or third time that sperm bank half siblings would meet. Tom and Alton would be inventing an entirely new relationship: although half siblings have existed for as long as men have been cheating dogs, the sperm bank brother was something new. Regular half siblings have a *known* father in common: They share a family history, a name, a life. But sperm bank half brothers have only DNA in common; their shared father is a complete blank. Coral was not a real person to Alton and Tom. They didn't even know his name. The only thing they knew about him was that they didn't know anything about him.

That paternal void was not the only obstacle. I also feared circumstances—nurture—would make it tough for the two boys to get close. By the time I put them in touch, Tom was sixteen and a rising junior in a decent public high school, Alton a rising freshman in a superb one. Tom lived in a middle-class house in a run-of-the-mill midwestern suburb, Alton in a beautiful house in the heart of Cambridge. Tom whiled away his time on the usual pursuits of the teenage boy—video games, wrestling, girls, rap—while Alton was a serious student musician and artist. Both moms were strivers, but they lived in different worlds. Tom's mom had battled her way to a bachelor's degree and a job in technical support; Alton's had earned graduate degrees at the best universities in the world and was one of the top women in her field. Was mere blood thicker than these differences in background, temperament, interests, and income?

Tom was over the moon when I told him about Alton. He

couldn't believe he had a brother already. Once the mothers gave permission, Tom fired off a chatty, boisterous e-mail to his new brother, Alton:

> Hi! I dont have much of an idea what to tell you about me. So ill just tell you anything that pops into my head. Right now im in a band named infernal. We're a rap group. OK stop laughing now. I have a lot of fun making music and my friends think I am the best person in the group at being able to "catch" the beat. College has been an important part of my life. Also I have about 19 credits at college and a 3.8 gpa. The classes I have taken include English 1, English 2, Psychology, American Government, Principles of Microeconomics and a bunch of computer classes . . . I am also spending time with my girlfriend Lana. Shes a really nice girl from Russia. Russian is her first language but you can barely tell when your just talking to her cause she doesnt really have an accent, but when I go to her house though it is really weird like hearing her give her dog commands in Russian. . . . Um what else . . . I have three cats and one of them me and my friend Mike found wandering outside my house (the poor kitty only had three legs so we had to take it in), no we didnt call it tripod. Well thats about all that I can think of to tell about me right now, ask me some questions in your reply so I will have a better idea in what your interested in finding out about me.

Alton answered cheerfully:

> I'm not sure where to begin so I guess I'll just start in a random place and go from there. I'm 15. . . . I play the piano, though not as seriously now, and if you have an mp3 player

(I hope) you can hear me play a solo piece in Italy . . . I know
I was rushing but it was the last night in an 8 concert series
and it was late at night and the room was hot with a bunch
of sweaty old italian ladies and their expensive perfume.

I like computer graphics a lot. Ok, I admit it, I'm a total
comp nerd, but who isn't nowadays. . . . I also have a pet but
mines a dog. Her name is, well wuddya know, it's Lana. She
is a four legged golden retriever and I think I have a picture
of her on my website. That's all for now, it's great to talk to
ya.

Tom responded eagerly, confessionally:

Actually I don't have an Mp3 player right now cause I killed
my other computer on accident. . . . So yeah as soon as I get
my computer fixed ill listen to your music. It sounds pretty
cool though. Im not sure what else to tell you about me. I
guess I should tell you right off the bat so it doesnt come as
a surprise later, but right now im stuck going to group therapy. Why you ask, because last year when I was in school
they found "suicidal" lyrics in my bookbag to a song I was
writing. What they didnt know and didnt care to know is that
I was trying to write a song against suicide becayse that was
about a week after my friend Eden tried to commit suicide
by swallowing about 50 tylenol and we wouldn't have found
out if she hadnt collapsed in front of us . . . well anyways
that had a profound effect on me seeing one of my close
friends in the hospital because she tried to commit suicide,
so I was writing a song about it when the school found it, so
I ended up getting suspended and I have to go to therapy
right now. I have found that most of my friends dont care
about it cause most of them believe me that I was writing a

song about how it is a bad thing to do that, but I have some friends who wont hang around me now because they think I am going to hurt them or myself. So I thought I would tell you, so it wouldn't come to a shock later. That was a depressing song when we got it finished, most of my songs aren't like that though. I usually write songs that make people laugh. . . . My band's website is still under construction since my friend mike basically wrote the site and I just did the HTML coding and since every other word he says is a cuss word its got a lot of cussing in it. I mean this guy uses about six cuss words to describe a newborn puppy. Hes my best friend though so I cant complain too much. I thought I would warn you though that there is a good amount of cussing on the site though so that when we get it up and I give you the address (if you want it) you wont be surprised and offended . . . we decided to put up a website so we could take online orders for our CDs, cause last year we released the cd and sold about 70 copies to people, most of whom don't even like rap music and found out there were still a lot of people who said that if they had heard a sample of our music and liked it they would have bought the CD. , , , , wow ive been talking wwwaaaaaayyyyyy too much, you probably stopped reading about 1 page ago :-). Well anyways write me back when you get a chance.

—Tom.

Alton wrote back:

That's rough about your friend, I'm sorry. How long do you have to be in therapy? The only therapy I ever did was when my parents broke up and all the dude did was nod his head and say, "interesting" or "hmm." Oh well. But think of it this

way: those people who won't hang out now are the one's
who wouldn't stick by their friend, so I guess they would not
be a true friend anyway.

That's cool about your band though, that you're selling
CDs . . . I can't rap or play guitar though, so I could never do
that. I think my mom sent a picture of me to you, at least
she said she did . . . did you see it and do you look at all like
me? That would be very cool.

Anyway I have to go, so write me back . . .

Alton.

Samantha, who had been hearing about the e-mails from Alton,
wondered if the two boys were brothers at all. They seemed so
different: Insane Clown Posse versus classical piano, suicidal friend
versus golden retriever. The two moms exchanged photographs. Mary
thought the boys looked very similar; Samantha thought they didn't.
(As the independent arbiter, I agreed with both of them. At first
glance, they didn't look much alike, but they share extremely deep-set
blue eyes, a wickedly strong chin, light brown hair, and a unibrow.
And there was something—maybe a look, a cant of the head—that
made them look like brothers.) Mary assured Samantha that Hazel
had checked the records: Coral was definitely Tom's dad, too.

Tom was in great spirits. He had a father, even though he hadn't
found him yet. He knew his real dad was a brilliant man (160 IQ!), a
family man, and an accomplished man. For the first time in his life,
Tom knew what it felt like to be proud of his father. Even better,
Alton seemed as if he were becoming a friend and maybe even a real
brother.

There was even better news, but Tom didn't know it yet. I was
keeping a secret from him, a secret of Samantha's. One day while we
had been talking, Samantha had confessed, "I know Coral's first
name. And I know what he does."

"Excuse me?" I said. Samantha said she had not told me everything about her 1985 visit to the Repository. Julianna McKillop had not merely shown her a photograph of Coral, she had revealed Coral's first name and profession. "When I was holding the picture of Coral in my lap, Julianna said, 'Oh, you better give me back the picture of Jeremy.' Julianna was incredibly embarrassed that she had said his name. She also told me Jeremy was a doctor, and I asked what kind, and she said surgeon. She told me that Jeremy had been in practice near Miami. She told me that Jeremy was open about being a donor, that he had gone on a TV program to talk about it. And she told me that Jeremy would be happy to meet my son when he was a teenager."

This wasn't the only news, Samantha said. She admitted that she had been searching for Jeremy/Coral and thought she had found him. When she'd heard the Repository was closing in 1999, she'd written a letter to the director. She'd told him she had an implied contract with the Repository, that Julianna had agreed that the Repository would contact Donor Coral when Alton was a teenager. The director wrote back and said: Tough luck, that's against the rules.

So Samantha had decided to find him herself. She was friends with several adoptees, and she had seen how they always wanted to find their birth parents eventually. She assumed Alton would want to know Coral one day. She had figured she'd better start looking now, while the sperm trail was warm.

Without telling her then-husband or her son—who didn't yet know he was a sperm bank baby—Samantha had begun trolling the Internet for Dr. Jeremys. She combed through lists of surgeons in Miami, Fort Lauderdale, Tampa, looking for Jeremys, Jeremiahs, and Jerrys. Whenever she found one, she would send him an e-mail saying, "I am looking for a sperm donor named Jeremy who donated to the Repository for Germinal Choice. If it is not you, do you know who it is?" Most never wrote back. Some wrote back saying, "It's not me, and I don't know who it is."

After a year and dozens of failed letters, she uncovered a lead. In late 2000, a few months before she saw my article on *Slate*, she got an unusual reply to one of her letters. A Dr. Jeremy wrote back, saying, "It's not me, but I think I know who it is: Dr. Jeremy H. Taft." Samantha followed the clue. She surreptitiously checked out Jeremy H. Taft. He was a celebrated, talented Miami plastic surgeon, and he was a perfect match for Donor Coral. He had an interest in math, significant musical ability, and the correct number of kids of his own (three). He had written a book, just like Donor Coral. Still, Samantha was ambivalent, because Jeremy H. Taft was sleazy. He had a huge practice, largely because he advertised his services on city buses, on billboards, and in magazines. On the other hand, Samantha discovered he ran a scholarship program for needy kids.

Whatever he was, he was definitely the right guy. He was the right age, had the right hair and eye color, the right marital status. He even looked like the picture of Coral she had seen. His personality fit. A man who would erect a billboard of himself advertising face-lifts was the kind of man who would go on television to brag about donating sperm.

Samantha told me she had written Taft two letters and so far received no reply. But she wasn't sure she had sent them to the right address, and she wondered if the letters had been vague. She had also left a couple of phone messages at his office that had gone unreturned, but again, she suspected they had been too opaque. He might not have understood the communications because she'd never said straight out that she believed he was Donor Coral and had plenty of supporting evidence.

Samantha revealed this to me right at the time Alton and Tom started corresponding. She was sure Jeremy Taft was Coral, and she believed he would be glad to hear from her and Alton, as long as he was sure they were legit and not seeking money. So how could she reach him? She decided to write a much bolder, clearer letter. She would present all her evidence in its most conclusive form. She would also tug at whatever paternal feelings he had by telling him all about Alton. She would enclose a photograph: What man could resist a photo of his own handsome son? She drafted this letter:

Dear Jeremy,

I believe that you are the genetic father of my son Alton. He was conceived via a sperm donation from the Repository for Germinal Choice, and was born August 19, 1986. Julianna (from the Repository) slipped up and told me the donor's first name when she showed me his photo in 1985. She also told me he was a surgeon in Florida and told me about his sister's musical gifts. I knew of other attributes from the Repository's description. It was not hard to find you with help from the internet.

Alton is quite an amazing and wonderful kid who would make you proud if you knew him.

He understands full well the fine balance between hard work and creativity. He is soft spoken and modest, with a sweet and happy personality.

His hobbies are computer games and mountain biking. He likes chess and nonfiction writing.

If you have any interest in contacting him, his e-mail address is _____. If you are at all interested in meeting him, I could arrange to bring him to Miami. We are now a tiny family of two and it could be quite wonderful for him to meet you or other members of your family, especially your parents or children. On the other hand, just a photo from you could become a cherished possession.

Samantha Grant

She attached a cute photo of Alton, sealed the letter and picture in an envelope marked "Personal," placed that envelope inside another envelope, and mailed it to his office. No way could he duck it.

In the meantime, Samantha and I discussed whether to tell Tom about her discovery of Jeremy Taft/Donor Coral. We thought we should keep it a secret for now. What if Jeremy wanted to meet only one child? It would be cruel to tell Tom that his father was out there but wouldn't see him. And it would be unfair to Jeremy Taft to saddle him with two sons when he knew about only one. Besides, Tom was just sixteen; was he old enough to handle the information about Jeremy and the frustration of knowing he might never get to meet him? We agreed: until Jeremy Taft told her he wanted to meet Tom, we wouldn't tell Tom or Mary about him.

Still, I felt guilty. There was one thing in the world that Tom wanted—to know his father—and I was depriving him of it.

CHAPTER 5

DONOR WHITE

Donor White's entry in the 1988 Repository for Germinal Choice catalog.

The same week Samantha Grant left me her cryptic, anonymous voice mail, another mysterious woman called late at night. "I'm a mother of a ten-year-old girl from the Repository," she whispered into my voice mail. "I want to talk to you." Click.

I lingered by my phone the next day. No call back. The following morning, the phone rang, and I heard the same whispering voice. "I called you the other night. I have a ten-year-old daughter who was born with the help of the Repository."

I said hello and asked, "What's your name?"

She didn't answer for a moment, then whispered, "Beth Kent. My daughter is Joy." I cleared my throat and summoned up a whisper of my own. I never really had to whisper until I started working on this sperm bank story, but there is something about these conversations that makes people talk softly. It's not the need for secrecy; they would whisper to me when they were alone in their houses. It was more like the hush of the confessional, the sense that these are sacred matters—birth, identity, genius—not to be joked about or even discussed in everyday voices. Even the e-mails from mothers and donors felt like whispers. No exclamation points. No smiley faces.

I asked Beth why she called. She said she wanted to dispel the notion that the women who went to the genius sperm bank were crazies seeking *über*-children. She told me she had gone to the Repository not because she wanted a genius baby but because she wanted a *healthy* one. The Repository was the only bank that would tell her the donor's health history. She had picked Donor White. Her daughter Joy, she said, was just what she had hoped for, a healthy, sweet, warm little girl. (That's why Beth asked me to call her daughter "Joy.") "My daughter is not a little Nazi. She's just a lovely, happy girl." She described Joy to me, how she loved horseback riding and Harry Potter. She read me a note from Joy's teacher: "Wow, it is a pleasure to have her smiling face and interest in the classroom."

Even so, Beth was cautious. She was suspicious of reporters. I sensed that she was feeling me out in the conversation, gauging whether I seemed to exploit her. At some point, she must have decided that I passed the test, because she said in a new, confiding tone of voice, "Now I want to tell you a little bit more about what happened to us. Maybe you can help us. I was hoping that you might be able to help me find Donor White. I'm looking for him, and I am pretty sure he is looking for us, too."

Oh, boy, here came another secret. I had not been privy to this many secrets since the late-night drunken spill sessions in college, and I was starting to feel funny about it. I don't think I am a particularly good confidant. I can keep a secret well enough, but I don't have the instinct for knowing the right thing to say when I'm told it. Still, I was starting to realize that empathetic ability was not so important, at least when it came to the Nobel sperm bank. I was the useful idiot here. Mothers were not confiding in me because they hoped I would go all Oprah on them and counsel them on how to live a better life. They were confiding in me for strictly pragmatic reasons—because I could help. Like a movie detective, I was being told secrets only because I might be able to unravel them.

Beth unfolded the mystery for me on that first day and over several weeks with phone calls, e-mails, and packages of letters. Eventually, she let me visit her and Joy in their small Pennsylvania town. I liked her a lot, especially once we dispensed with the solemn whispers. She was a nurse, which didn't surprise me a bit. She was at once brisk and warm. I could imagine her jabbing the needle, and I could imagine her giving the hug afterward.

Here's the story she told. In the late 1980s, Beth and her husband were living in San Diego. She was finishing up a stint in the coast guard. He was a fireman, a bit older. He'd had a vasectomy, but Beth was desperate for kids. She was in her early thirties and couldn't wait any longer. Beth was leery of sperm banks. She had inquired at several banks, but they had been closemouthed about who their donors were. How could Beth know if the guy was really healthy? What if he had a family history of mental illness or cancer or heart disease? She wouldn't take the risk. But in 1988, she read an article about the Nobel sperm bank. It sounded different. She sent away for an application and a catalog and was pleased to discover that the Repository told her everything she wanted to know about the donors: notably that they had passed blood tests for major illnesses, that their family medical history was clean, and that they had led lives of accomplishment, happiness, and good health. Beth knew nothing about Robert Graham's eugenic ambitions, and she would not have been interested if she had.

Beth filled out the Repository application, which she found gratifyingly comprehensive. She liked that she had to give her medical history and that she had to promise not to smoke or drink while pregnant. The bank's Escondido office wasn't far from Beth's home. After mailing back the application, Beth and her husband dropped by to introduce themselves to Dora Vaux, then the Repository's office manager, recruiter, and fairy godmother. They asked her advice about which donor to choose. Dora studied Beth's husband for a minute, then declared, "Donor Turquoise or Donor White." Dora flipped open the catalog to Donor Turquoise #38, who was described as "a top science professor at a major university, head of a large research lab. Has published several college text books." His IQ was 145. He was of German descent and had blue eyes, fair skin, and thick, wavy brown hair. He was born in the 1940s and was "very outgoing, happy, confident." His hobbies were "farming, drama, German literature." He had two "healthy, bright children," and "all his family are very long-lived."

Then she turned to Donor White #6: "Scientist involved in sophisticated research. Many technical publications." He was of English ancestry, with blue eyes and dark brown hair. He was taller than Turquoise—six feet to Turquoise's five feet, eight inches—and a decade older. His personality was "very engaging, warm, friendly." He liked running, gardening, and reading history. His family was long-lived, too. His mother suffered from migraine headaches; he was nearsighted.

Shopping for a father! That special Turquoise glow or White's whiter than whites? Crest or Colgate? Beth agonized. Dora asked

her what she *really* wanted. Beth said, "I want to have a happy baby."
Dora told Beth that Donor White was a friend of hers, he lived
nearby, and he was a delightful, sweet man. She told Beth that
Donor White had already fathered lots of children through the
Repository, and they were all happy, happy babies.

Then Dora dug into her file cabinet and pulled out a copy of an
essay, "The First of My 12 Children Will Soon Be Four" by "R. White."
Donor White had written it, Dora said. Beth took it home and read it.
Donor White described how he and his wife had been childless
through more than twenty-five years of marriage, because of unknown
fertility problems. But when he was over fifty, he'd been recruited as a
sperm donor at the Repository and had been amazed to discover that
he was very fertile. Now a dozen babies had been born from his seed in
only three years. Donor White wrote how happy the children made
him feel, even though they were unknown and unknowable:

> The indirect success described above is not like having your
> own children, of course, and I will likely never be able to see
> any of them in person. . . . Moreover, many of these chil-
> dren will likely never know that their adopted fathers are not
> their biological fathers. Still, I know these children are out
> there somewhere, and they are thought about often. I have
> seen very pleasing photographs of several of them, with their
> parents' permission, and have been able to form my own
> mental images of others while running on the beach in the
> quietness of the early morning. This is a rather poor substi-
> tute for having one's own children, but it does provide a
> sense of continuity that was not present before. In my view,
> a person's genes really belong to all of those many ancestors
> from whence they came, and we are only allowed to borrow
> and make use of them during our lifetimes. I have the satis-
> faction, then, of having been able, in an anonymous way, to

connect the past with the future in a continuous line like a curve on a graph.

The article touched Beth. She *liked* Donor White. She ordered several vials of White sperm and scheduled inseminations with her doctor. Month after month passed, and she couldn't get pregnant. Beth occasionally visited the Repository office to cheer herself up. Dora and Robert Graham had covered the office walls with photographs of the sperm bank's offspring. These beautiful babies frustrated and enticed her. She ached for a child. Dora hugged her, comforted her, and told her it would all work out.

Beth was so desperate to conceive that she quit her job for one with better health insurance. After six months of failure, she gave up on regular insemination. She spent almost all her savings on in vitro fertilization, trying to have a test-tube baby with Donor White's sperm. This was 1989, when IVF success rates were very low and the cost was very high. But the pregnancy took.

Beth swore she wouldn't raise a kid in Southern California—too dirty, too dangerous. She and her husband moved back east to Pennsylvania farm country. Beth had grown up in small-town Indiana. She desired the same kind of childhood for her daughter. Joy was born in the summer of 1990, a boisterous, darling little blondie. After all her fertility trials, Beth considered Joy a miracle baby. Her thoughts returned again and again to Donor White. She thought about his article, and about the children of his own he could never have. When Joy was a few months old, Beth wrote him a thank-you note, care of Dora at the Repository, and enclosed a picture of tiny Joy.

Then, when Joy was almost a year old, Beth had a more radical idea. She was traveling to southern California on family business. She called her old friend Dora at the Repository. Beth offered to leave Joy with Dora at the Repository for a couple of hours, in case Donor White wanted to come by and meet his daughter. Now, at any other sperm bank in the country, the office manager would have declined that offer politely. Allowing anonymous donors to meet their children violates every principle of sperm bank confidentiality. It sabotages the contract among donor, parent, and bank. American sperm banking is premised on the notion that the donor is anonymous and free of all the obligations (and pleasures) of fatherhood. Allowing visits could open bizarre legal trapdoors. What if the mother dropped off the child and never returned, for example? The emotional bonds created by such meetings would be tragically precarious, as fathers bonded with children that could never really be theirs.

But Dora's attitude toward rules was erratic. (After all, she had told Lorraine the full name of Donor Fuchsia.) And Dora had a warm heart; Donor White was her friend. He lived around the corner. Besides, no one but Beth and Donor White would ever know the meeting had occurred. So Dora said yes, come on by and I'll babysit Joy.

So, on the afternoon of June 2, 1991, Beth dropped Joy off with Dora at the Repository's small office in Escondido. Beth headed out to a coffee shop, telling Dora she would return in two hours. Dora called Donor White. He and his wife raced over from their house.

When Beth came back, Joy was clutching a gift from Donor White, a Playskool doll wearing a pink dress. Dora told Beth that Donor White had been ecstatic to meet his daughter. "He told Dora that he would live on that moment for the rest of his life."

After they returned home to Pennsylvania, Beth locked the doll away in a keepsake chest. As Joy grew up into a toddler and then a little girl, Beth would occasionally bring out the doll and say, "Someone special gave this to you."

Her gratitude to Donor White endured. Beth would mail photos of Joy to the Repository almost every year, always enclosing an extra print and asking that it be sent to Donor White. Then, right before Joy's fifth birthday in 1995, Beth and her husband received an envelope from the Repository. Inside they found a birthday card for Joy. It was signed "Donor White." There was nothing else on the card.

Soon, a three-legged correspondence arose among Beth, Donor White, and the Repository. Beth would send a letter to Donor White, care of the Repository. The new Repository manager, Anita Neff, would cross out all identifying information and forward it to Donor White. He would send a reply to Anita, who would edit it and mail it on to Beth.

In December 1995, Beth mailed the Repository a Christmas card for Donor White, along with a photo of Joy and Santa Claus. The day after Christmas, Donor White and his wife replied with a long letter. Donor White described some of his favorite ancestors—without naming names, of course. He discussed his fascination with DNA research and his hope that Joy might become a microbiologist, before adding wryly, "We won't make Joy select a career before she finishes first grade."

Donor White wrote that he and his wife hoped they would meet his daughter, even though he knew it was impossible. "In the back of our mind there is the thought that some day, some way, we might get to make a future visit in person. In the meantime, please know you are thought of very often, Joy, and thank you for letting us believe that we really do have a small part in your life." The letter was signed, "With all our love, Your adoptive grandparents."

The next summer, Beth mailed Donor White a large photograph of Joy skiing in a red snowsuit and a videotape of Joy's ballet recital. Anita Neff rejected the videotape—Joy was too identifiable—but forwarded the skiing picture because Joy's face was sufficiently blurry. Donor White replied with a letter announcing that he had framed the skiing picture and mounted it in his living room. On Christmas Day, 1996, he composed a poem to Joy inspired by the photo. He called it "A Figure in Red on a Field of White": ". . . May your path through life be as smooth and happy as on that day, when over the snow you did joyfully glide away. . . ." He sent it to Beth. That was the last she heard from him.

Three weeks later, Beth received a letter from Anita Neff. The Repository's board of directors was cutting the correspondence, Anita wrote. "We simply cannot continue to share Joy with the donor. A unanimous decision was made to discontinue any further interaction between donor and offspring as it breaks the rule of confidentiality. While this has been the rule of the Repository all along, we recognize it has been bent for you in the past. . . . However, no further interaction will be allowed."

Beth was disappointed. The letter arrived just as her marriage was breaking apart. Her father had died soon before Joy's birth, and her mother was sick. She and Joy were practically alone in the world. Beth had heard of women who fell in love with their unknown sperm donors, fantasized about them as real husbands and fathers. That wasn't her. She had no romantic dreams about Donor White. He was married and way too old for her. But she had hoped he would be the steady grandfather that Joy had never had.

Beth puzzled over how to find Donor White. Dora Vaux, who might have helped her locate him, had moved away. (Dora died before Beth ever thought to look for her.) Beth reviewed what she knew about Donor White: that he was about sixty, that he had lived in California, that he liked to run, that he had a much younger sister and a niece of college age. She surmised, from hints in his letters, that he might work for the Human Genome Project. She wondered if his last name might in fact be White, that it wasn't merely his donor ID. She starting hunting, slowly, for biologists named White who lived in the West.

She didn't get anywhere, and soon, she stopped looking. Life took over. Joy was growing up from a bouncy baby into an enthusiastic girl. She had the energy of eight kids. She played soccer and basketball and rode horses. She took harpsichord lessons. Mostly she danced ballet: *Nutcracker* season was the highlight of the year. The older Joy grew, the more Beth believed in nature over nurture. Joy was extroverted whereas Beth was quiet; that had to come from Donor White, she thought.

Beth, now remarried, worked as a nurse but designed her life around Joy's. She worked only during school hours. In the afternoons, she shuttled Joy from game to practice to rehearsal. Beth was proud of her daughter, but in a measured way. Joy was not an egghead, Beth thought, but she was a bright girl and a good student. She was healthy, athletic, and friendly. Whenever Beth thought about the Repository, which wasn't a lot, she judged that it had given Joy a small helping hand: Beth had expected her daughter to be sweet and smart and pretty, but Joy was a little sweeter and smarter and prettier than she'd expected.

Beth knew, rationally, that she was not entitled to meet Donor White. That had never been part of the deal. She had happily signed the Repository's confidentiality contract. At the time, she had never anticipated wanting to break it. She had believed her marriage would last, that her husband would be Joy's father in both name and action. And even if she had wanted to break the contract, she understood its necessity. It would be a mess if donors and kids were always hunting for each other. What if a kid found a donor who didn't want to be a father? What if a donor found children whose own father objected to the donor meeting them?

But the makers of the rules had never imagined the possibility of a Donor White, Beth thought. The Repository had violated its own regulations: It had let her correspond with him. It had let him meet Joy. It had encouraged the relationship among father and mother and daughter to blossom. Just because it might be a mess if *all* donors and children could connect, that didn't mean it was wrong for one family. Beth knew that Donor White was a good man, that he was no threat to her daughter. In this one case, surely, the system could bend.

In 1999, two years after she lost contact with Donor White, Beth heard that the Repository had closed. She figured that her last chance to find Donor White had vanished. Beth decided she needed to tell Joy about Donor White. Beth had seen too much loss in the hospitals where she worked, too many people who died young and left things unsaid. She didn't want to leave anything unsaid to Joy. So one day in 2000, Beth explained it all to Joy in a way that a ten-year-old could understand. She told Joy that her daddy was her daddy, but that she *also* had a "donor daddy," a special, very smart man who had helped her get born. Beth read Joy one of Donor White's letters. She pulled out the Playskool doll again and told Joy where it had really come from. Joy might never meet the donor daddy, Beth said, but she should know that he thought about her every morning when he ran on the beach. Joy wasn't surprised to learn she had a second father. "She loves her dad, but he is very different from her," Beth told me. "I think it made sense to her that there could be this other father, too."

Joy had just read the Harry Potter books. She told her mom that she thought Donor White was like Dumbledore, the kindly old wizard. Joy was curious to meet Donor White, but she was ten years old and busy. She didn't think about it much.

A year after the conversation, Beth happened across one of my *Slate* articles. She realized she had one last chance to find Donor White, and she called me.

I agreed to help her. I published "A Mother Searches for 'Donor White'" in *Slate* on February 27, 2001. In the article, Beth said she wanted Donor White to know that Joy was a sweet and enthusiastic girl. That she was "kind of competitive." That she played soccer and "is all over the field." That she was very pretty. That she "does well in all of her subjects, but social science interests her most." That teachers liked her but she also had lots of friends. That she was taking riding lessons. That "she puts her heart into life."

At the end of the article, I listed my e-mail address and phone number and invited Donor White or any other families that had used Donor White sperm to contact me. (Beth was also hoping to find siblings for Joy.) I had never been so certain in my journalistic life. I was sure Donor White would stumble across it or that a friend who knew his secret would see it and tell him. At the very least, I knew that another Donor White family would contact me and give Joy a brother or sister. It couldn't fail—there was too much karma on its side: a darling eleven-year-old girl, hoping for a grandpa!

I have never received such a response to a story. E-mails poured in from around the world: other sperm bank kids wishing for Donor White as a father, other sperm donors praying that their children were searching for them, other mothers heartbroken for Beth and Joy. TV requests piled up in my in-box—*Primetime Live, Unsolved Mysteries*, and countless documentary makers begged Beth and Joy to go on the air. Donor White would be sure to see them, the producers promised.

Beth and I were both optimistic. But the months slipped by, and we heard nothing. I called Beth. She was so frustrated that she was considering doing one of the TV shows. She thought television might be a better way to reach him than the Web. Donor White is in his late sixties by now, she said, and maybe he doesn't use the Internet.

Or maybe he's dead, I thought.

CHAPTER 6

BIRTH OF A NOBEL
SPERM BANK CELEBRITY

Robert Graham strikes his favorite pose. *Eric Myer*

February 29, 1980, the day of the Repository for Germinal Choice's public debut, was a triumph for Robert Graham—a nation enthralled, Nobelists enlisted to the cause, marvelous sperm chilling in the freezer, Mensa women lining up to acquire it. That was the Nobel sperm bank's first great day, and its last one. Disaster struck immediately, in the person of William Shockley.

By admitting to the *Los Angeles Times* that he had donated, Nobel Laureate Shockley had rescued Graham from humiliation.

He had proved that though Graham might be a kook, he wasn't a fraud. Later Graham would say that he was "eternally indebted" to Shockley: "He was the one person who saved me from looking like the country's champion liar." But in saving Graham, Shockley condemned the sperm banker to a greater disgrace.

We live in William Shockley's world. As a scientist, he invented the transistor and lit the fuse on the information age. Before Shockley, the radio was elephantine, manned space flight was fantasy, computers were preposterously large, clunky, and expensive. His transistor changed everything. Later, as a businessman, Shockley midwifed the birth of Silicon Valley and kicked off the greatest commercial revolution in American history. He brought the country's most promising young scientists to California and set them to work solving practical problems. Shockley's shock troops would go on to create the most important technology companies in the world. For these two achievements, Shockley deserves the world's gratitude. But as a—who knows what, really: philosopher? gadfly? racist?—Shockley squandered that goodwill and made himself one of the memorably noxious public figures of the twentieth century.

Born in 1910, Shockley was the only child of a wealthy old British mining engineer and his gutsy young American wife, May Shockley. May had graduated from Stanford and then worked as a mineral surveyor in roughneck Utah mining camps. May transmitted her ambitious vigor to her son. She indulged him, worshiped him, dominated him. From his infancy, she told him, "Bill, go set the world on fire." Shockley was a cocky baby who became a cocky man. Pictures show a masterful child, his arms crossed, his legs akimbo like a tiny colossus. His parents couldn't control him; kids in his Palo Alto neighborhood couldn't stand him; but he was dazzlingly smart. (Though when Stanford professor Lewis Terman recruited for his famous study of gifted children, Shockley didn't make the cut. His 129 IQ was too low, a slight that Shockley carried with him to the grave.)

Shockley's father died when he was a teenager, an event that seems to have left no impression on him at all. Moving to Los Angeles, on the other hand, did. Shockley attended Hollywood High at the height of the Jazz Age. The blossoming movie industry entranced him. Shockley cultivated a cinematic conception of himself, a kind of posturing that stuck with him throughout his life. According to one friend quoted in Michael Riordan and Lillian Hoddeson's excellent history of the transistor, *Crystal Fire*, Shockley styled himself as a cross between "Douglas Fairbanks Sr. and Bulldog Drummond, with perhaps a dash of Ronald Colman." He swashbuckled his way through high school and college, driving a stylish De Soto roadster and packing a pistol—for no reason other than to have it. Shockley treated every social encounter as a chance to show off, demonstrate his intellectual superiority, razzle-dazzle boys with a magic trick, flirt with girls.

He had the brains to match the style. He raced through Caltech studying physics, then, inspired by quantum mechanics, headed east to earn a speedy doctorate at MIT in 1936. When Bell Labs, the world's greatest institution of practical science, lifted its Depression-era hiring freeze, Shockley was the first man it called. It assigned Shockley to improve its vacuum tubes, the expensive, cumbersome backbone of Ma Bell's phone network.

World War II interrupted. Shockley soon demonstrated the ruthless thoroughness that would define his career. Shockley believed that anything—a physics conundrum, a conversation, even a person—could be reconfigured as a series of logical problems. When he bought a boat, for example, he devised logical rules for steering it—a system of scientific sailing. He knew that unfettered rationality would always find the right answer. Dispatched as an adviser to the Navy, Shockley was appalled at its haphazard methods of hunting German U-boats. So Shockley devised an algorithm that instructed when, how, and how often to drop depth charges. The

Americans' U-boat kill ratio jumped sevenfold. Later, Shockley devised a new formula for targeting and dropping high-altitude bombs. The Navy awarded him its highest civilian prize. Decades later, when his critics tarred him as a Nazi-sympathizing white supremacist, Shockley gleefully reminded them of his medal.

After the war, Shockley took command of Bell Labs' effort to build a "solid-state" amplifier, a device that could amplify signals like a vacuum tube but be smaller, cheaper, cooler, faster, and more reliable. In late 1947, two of his subordinates, Walter Brattain and John Bardeen, built the first working transistor, a piece of germanium (a close cousin of silicon) attached to two small wires. It could amplify a signal like a vacuum tube did—one hundred times as much electrical power came out of the first transistor as went in—but required a fraction of the space and energy.

Shockley grabbed the credit for his underlings' work, but secretly he was infuriated that Bardeen and Brattain had made the discovery without his help. Over New Year's, Shockley locked himself in a room in Chicago's Bismarck Hotel and spent three days figuring out how to improve Bardeen and Brattain's invention. Shockley devised a better device called a "junction transistor." His junction transistor would be infinitely easier to manufacture and use. His transistor—not the "point-contact transistor" of Bardeen and Brattain—would eventually revolutionize electronics, computers, and practically everything else in the world.

In June 1948, Bell Labs announced the transistor, but hardly anyone noticed. Shockley's colleagues and other physicists considered the transistor to be simply a one-for-one substitute for vacuum tubes. It would do little but shrink radios and telephone networks. Shockley was practically the only person who understood its significance. He predicted, with astonishing insight, that the transistor would become the "ideal nerve cell" of small, vastly more powerful computers. It took decades for the rest of the world to recognize what Shockley knew immediately: that this was the most important invention of his lifetime. (Bill Gates has said, "My first stop on any time travel expedition would be Bell Labs in December 1947.")

By the early 1950s, Shockley's junction transistors had graduated from obscurity to the mass market, manufactured by the thousands and millions. (They were made of silicon, which worked better than germanium.) Portable transistor radios flooded into the United States from Japan. *Fortune* magazine declared 1953 the Year of the Transistor.

But Shockley seethed. He couldn't profit from his invention, because Bell Labs owned its employees' patents. His showmanship and credit hogging alienated Bardeen, Brattain, and his other Bell Labs colleagues. So in 1955, Shockley quit Bell, moved home to Palo Alto, and started his own company, Shockley Semiconductor.

Shockley brought the silicon to Silicon Valley. His arrival rang the opening bell for the Valley's boom, what Michael Lewis has called "the greatest legal creation of wealth in the history of the planet." Housed in a Quonset hut in Mountain View, Shockley Semiconductor wasn't the first high-tech company in the area— Hewlett-Packard beat it by sixteen years—but it introduced the transistor technology that would make Silicon Valley great. Shockley planned to manufacture better transistors, but what was more important was how he intended to do it. In the age of the Organization Man, Shockley had unusual ideas: His company would be run by scientists, not businessmen. It would pay well for talent. It would work fast. It would prize creativity over routine competence, and everyone would judge, ruthlessly, the work of his peers.

Shockley had an amazing eye for talent. His systematic brain had devised metrics for scientific productivity and he hired only applicants who scored high. He also ran his applicants through a battery of psychological tests. Shockley, one of the most famous scientists in the world, awed his young recruits: meeting him was "like talking to God," said Robert Noyce. Shockley hired a dozen of the best young minds in the country, including the amazing trio of Noyce, Gordon Moore, and Eugene Kleiner, and set up shop.

On November 1, 1956, a year after Shockley Semiconductor opened, Shockley, Bardeen, and Brattain were awarded the Nobel Prize in Physics "for their researches on semiconductors and their discovery of the transistor effect." It was one of the rare occasions when the Nobel board recognized a scientific breakthrough soon after it occurred.

The prize anointed Shockley as a symbol of American greatness. When *Life* magazine wanted to depict American scientific progress, it published the iconic photograph of Shockley, Bardeen, and Brattain demonstrating the first transistor. When Congress needed someone to discuss the future of American science post-*Sputnik*, it called Shockley.

Shockley loved his new celebrity. He was a born showman. He would hand out transistors to his audiences. He liked to do magic tricks at the podium, making bouquets of flowers appear and disappear as he spoke. He possessed the knack of explaining science with what Tom Wolfe described as "homely but shrewd" examples. This is Wolfe's account of Shockley on amplification: "If you take a bale of hay and tie it to the tail of a mule and then strike a match and set the bale of hay on fire, and if you then compare the energy expended shortly thereafter by the mule with the energy expended by yourself in striking the match, you will understand the concept of amplification."

Shockley, in a hint of what was to come, used his national pulpit to lobby for a scientific aristocracy. The American public was too democratic, divided, and stupid to make complex decisions, he said. Instead there should be an Institute of Public Enlightenment, a board of wise men that would objectively analyze complex issues and then use advertising to persuade Americans how to think about them. It was pure Shockley, populist marketing techniques to serve elitist goals.

As Shockley's public star ascended, his business collapsed. Shockley was a disastrously bad businessman and a worse manager. Shockley Semiconductor didn't manage to manufacture a single transistor. Shockley's employees detested him. After proposing to abolish hierarchy, Shockley built a dictatorship. He had promised his charges intellectual freedom, then bullied and abused them. In a hideously misconceived effort at openness, Shockley posted all salaries on a bulletin board, which embarrassed those who were highly paid and annoyed those who weren't. Shockley husbanded all the juiciest research for himself, hiding his results even from his employees.

Shockley's suspicion of his employees twisted into paranoia. When someone scratched a hand on a thumbtack, Shockley was convinced a plot was afoot. He forced some employees to take lie detector tests: Had they placed the guilty tack? Finally, on September 18, 1957, eight of his top scientists quit. In the original example of the entrepreneurial gumption and institutionalized disloyalty that now defines the Silicon Valley, they immediately founded a rival company, Fairchild Semiconductor. Shockley called the quitters the "Traitorous Eight," and he took his rage against them to the grave.

At Fairchild, the Traitorous Eight took Shockley's original but betrayed ideals and turned them into practice. Fairchild really did abolish hierarchy. Offices: gone, replaced by cubicles. Reserved parking spaces: gone. Dress code: gone. Anyone could make a decision; anyone could test a new idea. Their Anti-Shockley culture remade the American economy. At Fairchild, Noyce developed the integrated circuit, the method of linking transistors that made Shockley's imagined supercomputers a reality. Fairchild eventually fractured and seeded the valley. Noyce and Gordon Moore established Intel, where they created the most important hardware company in the world (and where Moore devised "Moore's Law," a conceptual framework for the computer revolution). Others of the Traitorous Eight built Teledyne, another Valley semiconductor shop. Eugene Kleiner founded Kleiner Perkins Caufield & Byers, the venture capital giant. Kleiner was the Johnny Appleseed of Silicon Valley culture. His VC investments spread the anti-Shockley ethos to virtually every important technology company of the past thirty years, including Sun Microsystems, Compaq, Amazon, and Google.

Shockley was the bastard father of the New Economy, but he himself foundered. His semiconductor company barely survived the Traitorous Eight's mutiny and limped along unproductively until the early 1960s. Shockley was in a funk. His own family didn't matter much to him; he had three children by the first of his two wives and mostly neglected them. His mother had ordered him to set the world on fire; he had done it once, but now the glory had dissipated and so he needed to do it again. The Nobel Prize honors its winners for "conferring the greatest benefit on all mankind." Shockley often cited this line as his inspiration. He took it not merely as a compliment but as an instruction. He would remake mankind, whether mankind wanted it or not.

Shockley was searching for a cause, and, in 1963, he found it. Ever since a wartime trip to India, Shockley had worried about overpopulation and vigorously supported birth control and abortion rights. In March 1963, Shockley saw a *San Francisco Chronicle* story entitled "Weird Attack Maims Shopkeeper." A hoodlum named Rudy "The Brute" Hoskins had burst into a San Francisco deli and tossed lye into the face of the owner, Harry Goldman, blinding him in one eye. It turned out that an ex-boyfriend of Goldman's girlfriend had promised Hoskins $400 to toss the acid (though paid him only $2). Shockley was mesmerized by the details of Hoskins's life: Hoskins was black. His "mentally dull" mother had an IQ of only 65. She had seventeen children but could remember the names of only nine of them. Shockley, rather than blaming the obvious villain— the creep ex-boyfriend who had paid for the attack—suddenly saw the world anew.

The problem was not that there were too many people on the planet, Shockley concluded. The problem was that there were too many of the *wrong* people on the planet (like Rudy Hoskins and his mom), and they were breeding too fast. With his systematic brain, Shockley immediately outlined the problem, named it "The Problem of Human Quality," and set about solving it. Shockley reached much the same conclusion that Robert Graham was reaching a few hundred miles south. The planet was being destroyed by "dysgenics." In better, bygone days, genetic failures like Rudy Hoskins's mother would have died before childbearing age. Now, thanks to what Shockley liked to call "humanitarianism gone berserk," society not only kept these imbeciles alive but paid them through welfare programs to have more children. Like Graham, Shockley discounted the fact that Americans were richer, healthier, and better educated than ever before.

It was no accident that Shockley developed his eugenic fixation at the same time as Graham. They were men cut from the same starched cloth. Like Graham, Shockley had been a giant in the 1950s, a scientific businessman in an age when there was no cooler job. Like Graham, Shockley had been fawned over as a wise man. And like Graham, Shockley started to fear the genetic decline of the U.S. population at the moment that Americans decided to stop listening to men like him.

Through the mid-1960s, Shockley laid out his dysgenic fears in articles and university lectures. In *U.S. News & World Report*, he wondered, "Is the Quality of the U.S. Population Declining?" Year after year, he petitioned the National Academy of Sciences to commission a study on American genetic decline.

Because he was a Nobel laureate, Shockley's theories about "human quality" were respectfully tolerated, but without great interest. He craved more attention, and he pushed until he found a way to make America really listen. He started talking more about race. Shockley had originally framed America's "quality" problems in generic terms: anyone could have a low IQ. But by the late 1960s, Shockley had narrowed his focus to the intellectual inadequacy of blacks. Shockley coined a phrase, "The Tragedy for the American Negro," that he repeated, mantralike, approximately every ten minutes. Black Americans, Shockley said, had an average IQ 12 points lower than whites'. As a result, few blacks could perform demanding jobs, and too many blacks were imbeciles. Social welfare programs and better education could not correct this problem; it was genetic. Blacks, Shockley said, were "genetically enslaved" to their poor DNA, condemned to lives of misery, poverty, and crime. These ideas about black intelligence didn't originate with Shockley; he built on the work of controversial social scientists—notably Arthur Jensen and Richard Herrnstein—who had been combing intelligence test data. Shockley viewed himself as a popularizer of their ideas. In fact, he was more of an unpopularizer. Their arid journal articles mostly escaped notice, but Shockley's inflammatory rhetoric, catchy phrases, and provocations enraged Americans.

Between 1960 and 1970, Shockley accomplished a remarkable reversal of reputation. At the start of the decade, Shockley was revered as a symbol of American greatness, one of the world's greatest scientific minds, a daring businessman, a hero. By decade's end, he was a national disgrace.

Shockley himself didn't seem like much of a provocateur. He discussed incendiary topics in a bizarre manner—exactly as if he were summarizing the latest advances in semiconductor research. He was the iceman. He didn't exude hatred for blacks—he didn't have any. He didn't exude sorrow—he didn't have any of that, either. Shockley's critics assumed that his racial anxiety stemmed from some personal experience, some deep trauma, but it probably didn't. He had no particular feelings for blacks one way or another. He hardly knew any blacks. To him, his racial conclusions were simply the logical outcome of a train of thought. As far as he was concerned, once he started to address human quality, he would follow its logic wherever it took him. In his mind, his conclusions had nothing to do with any actual black person; he was simply making an irrefutable point.

Shockley was at once a brilliant debater and a terrible one. Every point was backed up by a statistic, every sentence a model of logic, clarity, and chill. I listened to lots of tapes of him debating other experts, and he demolished his opponent every time. Yet his snooty, meticulous manner was exactly calibrated to infuriate anyone who disagreed with him.

Because Shockley was such a pure rationalist, he assumed that all problems were equally susceptible to his single kind of scrutiny. It never occurred to him that the "science" itself might be built on shaky premises. He believed religiously in the accuracy of IQ tests, that they measured intelligence in an absolute and total fashion. He never questioned whether an IQ test was a fitting tool of social policy. And he assumed that the "genetically enslaved" blacks led the lives of misery he ascribed to them, but he never actually talked to any of them.

Shockley chose precisely the wrong moment to share his conclusions with America. The late 1960s had arrived, with the full flower of the civil rights movement, the rise of black power, and the student revolts. Shockley was a comically delightful villain: the scientific racist. Tiny in stature, narrow-faced, mild of manner, he made a grand foil for blacks' anger and students' rage. He played along, touring campuses with the enthusiasm of a hot indie rock band. He would accept any invitation. He was shouted down by black students at Dartmouth; barred from Harvard and Yale; silenced, then whisked into a police car, at the University of Kansas. His supporters were attacked at California State University, Sacramento. His face was emblazoned on "Wanted: Dead or Alive" posters. When *Roots* aired, five universities canceled Shockley engagements.

At Stanford, where Shockley was a professor in the Engineering Department, students regularly demonstrated against him. They

William Shockley, the notorious Nobelist. *Courtesy of
Stanford University Archives*

shouted "Off Pig Shockley," pinned a list of demands to the university president's door with a hunting knife, and burned Shockley in
effigy. Shockley was unperturbed—and not-so-secretly delighted—
by the rage he generated. Black protestors dressed in white Klan
robes mau-maued his classroom. When his classroom was invaded,
Shockley serenely interviewed the disrupters to discern their com-
plaints, carefully chalked their demands on his blackboard, and
tried to debate each one point by point. He kept asking the protestors
"to formulate clearly defined questions." When a microphone at a
protest outside his office malfunctioned, Shockley cheerfully went
outside and fixed it so that the anti-Shockley imprecations could
continue. It surely satisfied him enormously to have all those transistors at work against him.

Shockley found ever more inciting ways to goad his enemies.
He praised Hitler's eugenic policies: "I would think that it would be

quite likely that there was some significant amount of elimination of genetic diseases [in Nazi Germany]. Just as the autobahns were a good thing, maybe there were some other good things about Hitler."

Shockley thought he could prove to blacks that whiteness led to intelligence. Shockley proposed to do this by measuring the percentage of "white" genes in blacks: he would show that the "whiter" the black person, the smarter he was. (Not that he had any real idea of how to test for "white" genes.) He asked NAACP leader Roger Wilkins to help him collect blood samples from members of the Congressional Black Caucus and other celebrated blacks, on the grounds that these accomplished people would surely prove to be significantly white. When Wilkins rejected him furiously, Shockley suggested that Stanford blood-test its five hundred black students. You can imagine how well that went over on campus.

In the late 1960s, Shockley floated his "Voluntary Sterilization Bonus Plan." The government, he said, should pay anyone with an IQ of less than 100 to be permanently sterilized—$1,000 for every IQ point under 100. (The cash would go into a trust, because such morons could not, of course, take care of the money themselves.) There would also be bonuses for bounty hunters who recruited willing candidates. At first Shockley called his sterilization plan an "intellectual exercise," but eventually he agitated to conduct a pilot program in California.

Naturally, Shockley beguiled Robert Graham. Shockley was not merely a Nobelist but one preaching what Graham himself believed. Graham contacted Shockley after his first controversial speech in 1965 and struck up a friendly correspondence with him. Graham sucked up to Shockley—he couldn't help it, Graham sucked up to all men he admired—donating generously to Shockley's Foundation for Research and Education on Eugenics and Dysgenics (FREED) and sending Shockley flattering "Dear Bill" notes.

So when Graham decided to start collecting his supersperm, Shockley topped his list. Shockley gave Graham his first Nobel sperm, supplying a sample during a 1977 visit to southern California. Shockley donated again in 1978. Soon after, Graham wrote a letter to Shockley telling him to expect more visits as soon as women started getting pregnant. "We will resume collecting donations as soon as we have begun to utilize a good proportion of our present library. We don't want such a superb asset to go long unutilized," Graham wrote. He signed off the note to Shockley, "I will keep you informed if you become a father again."

But Graham never used Shockley's "superb asset" again. Shockley, in fact, proved no asset at all, but Graham's biggest liability. When the *Los Angeles Times* announced the Nobel bank, scientists pelted the Repository with criticism. Geneticists objected that intelligence was not purely DNA-linked and hence Nobel sperm might not make kids smarter. Andrologists observed that Nobelists were too old to be effective sperm donors. Statisticians said Graham was duping customers who thought they were getting a guaranteed genius. But these slights didn't stick, since Graham's idea seemed harmless, at worst. If DNA did help intelligence, then these children could get a boost from a Nobelist dad. And if DNA didn't help intelligence, what harm was done?

The scientific criticisms didn't stick, but the criticism of Shockley did. The initial *Times* piece didn't make anything of Shockley's involvement. He was described only as a 1956 Nobelist in physics; his contentious second career as a racial scientist went unmentioned. But after a couple of days of favorable coverage for the Repository— or at least gape-mouthed coverage—newspapers around the country pounced on the Shockley link. Shockley turned the Nobel sperm bank from a curiosity into a menace and then into a joke. If Shockley was involved, the bank couldn't possibly be as innocent as Graham claimed. On his own, Graham seemed a little odd—an eccentric millionaire, yes, but well meaning and perhaps even visionary. Manacled to Shockley, Graham suddenly seemed sinister. Editorial pages in practically every major city denounced Shockley as a racist and wondered if the bank was a neo-Nazi plot. The *New York Post* head-line was typical: "Master Race Experiment."

A few days after Shockley's involvement was publicized, the widow of bank cofounder Hermann Muller wrote a letter to Graham demanding that he remove Muller from the bank's official name, the Hermann Muller Repository for Germinal Choice. Shockley's donation proved that the bank was a racist mistake, she said. *Time* and *Newsweek* mentioned the seventy-year-old Shockley's contribution to point out that sperm from older men was much more likely to produce a Down syndrome child.

Columnists reveled in the Shockley connection. San Francisco's Herb Caen said it was "proof that masturbation makes you crazy." Ellen Goodman mocked Shockley as the "Father of the Year." Every would-be comic from Art Buchwald on down riffed on the Nobel sperm bank: there were proposals for Academy Award sperm banks, media sperm banks, elite animal sperm banks, even sperm banks that were really banks. *The Boston Globe* fast-forwarded to 2003, when newspaper classifieds would advertise pedigrees and stud fees: "Nu-clear physicist, age 22, fee $750,000." Roald Dahl quickly pounded out a comic novel inspired by Graham. *My Uncle Oswald*, set in 1919, told the story of a temptress who seduced the world's greatest men—Renoir, Picasso, Freud, Einstein, Conrad, Shaw . . . —col-lected their sperm, and sold it to eager women. (Later, a Swedish novelist inspired by the Repository would imagine a similar sperm-collecting scheme at the Nobel Prize ceremonies in Stockholm.)

Saturday Night Live ridiculed Shockley and the bank in a skit entitled "Dr. Shockley's House of Sperm," starring Rodney Danger-field as himself. In the skit, Dangerfield was the most popular donor at "Shockley's House of Sperm"—a sperm bank whose wide selec-tion trumped that of "Jizz World" and "Jelly Barn." Clerks pleaded with customers to buy "some Linus Pauling"; they offered a sale on "the David Susskind." But all anyone wanted was Rodney Danger-field, Rodney Dangerfield, Rodney Dangerfield. Meanwhile, in the back room, Rodney was getting exhausted as order after order piled up: "Are you kidding? I can't. No way! No way! You kidding? It can't be done! . . . You're gonna kill the goose that laid the golden egg!"

Shockley adored the attention. It had been several years since he had made headlines, and he was delighted to be causing trouble again. Shockley, not Graham, became the public face of the sperm bank throughout 1980. Shockley, not Graham, was interviewed on *Good Morning America* and *Donahue*. (Shockley had the presence of mind to mock himself on *Donahue*. He stood up, all five feet, six inches of him, shed his suit jacket, and turned in a full circle in front of the audience. "It would be ridiculous for me to say I was a superman."

The bank and Shockley became inextricably linked in the public mind, and that proved disastrous for Graham. Despite his occasional flashes of wit, as on *Donahue*, Shockley mostly used his media appearances as occasions to appall his audiences. He spouted his racist theories of dysgenics and showed off charts of whites' and blacks' reproductive rates. He described the potential Nobel Prize babies as a "crop" and demonstrated his wonderful paternal instincts by expressing not a whiff of interest in the sperm bank children he might father. He cooperated with all journalists who called, to ill effect. He even sat for a profile in *Hustler* magazine. It began with memorable cruelty:

Jacking off for mankind! Somehow the scene comes off as comic, and maybe even a little sad: Imagine hand lotion or box of tissues at hand, and, of course, to excite the scientific imagination—a visual aid. A copy of *Hustler* perhaps. Confident in what he is doing, Shockley urgently pumps and pumps.

Outside a few feet distant, stands a California multimillionaire named Robert Graham. He is 73 and dressed in a business suit. With him is a white-jacketed technician, at the ready. They wait anxiously.

At length—success!

The door to the room opens. Shockley, who once opened the door to America's manned-space-flight program and a host of electronic marvels by helping to invent the transistor, shuffles out and hands over the warm specimen, a sticky splotch of semen deposited in a small plastic bag.

Playboy subjected Shockley to its famous interview in August 1980—Bo Derek was on the cover. In the course of it, Shockley admitted he hadn't known that sperm from older men carried a higher risk of birth defects. He also told *Playboy* about his disappointment in his own children, managing to insult his kids, humiliate his wife, and aggrandize himself, all at once. "In terms of my own capacities, my children represent a very significant regression. My first wife— their mother—had not as high an academic achievement standing as I had."

Shockley's return to notoriety didn't last long. He made a brief, and hilariously bad, run for the U.S. Senate in 1982 as the "Anti-Dysgenic Candidate," finishing seventh in the California Republican primary with only 7,000 votes. He embarked on bizarre projects, including commissioning a novel that was supposed to do for genetic decline what *Uncle Tom's Cabin* had done for slavery. (His author, a dentist's wife, wrote the first chapters of *Huntsville: A Journey into Darkness* before the project foundered.) He belittled his great physics research: dysgenics, he declared, was far more important than the transistor. His innate suspicion degenerated into paranoia. He taped every telephone call. He forced reporters to take a statistics quiz before he would agree to an interview.

Shockley, who had hobnobbed with kings and dazzled the world's greatest minds, ended in the gutter of American politics, allied with cranks and racists. During his senate campaign, the KKK offered to rally in Shockley's support and canceled only when Shockley's lawyer dissuaded them. Neo-Nazi newspapers took to reprinting Shockley's writings. When Shockley won a libel suit against an Atlanta columnist who had compared him to Hitler, the jury awarded him only $1 in damages: Shockley was such spoiled goods, the verdict seemed to say, that his reputation couldn't be damaged.

The Repository for Germinal Choice never received another vial of Shockley's sperm after 1978. This was a shame, a former Repository employee told me, because Shockley's sperm "quality"— if not his "human quality"—was very high. But after the *Playboy* interview, Shockley was spooked by the idea that his geriatric sperm might have health problems. He didn't want to donate until the Repository gave him a more thorough physical. It never happened.

The mocking and alarmist media coverage took its toll on Robert Graham and his sperm bank. When Shockley's involvement was publicized, protestors descended on Graham's estate, chanting outside his iron gates. Graham and his wife fled the protests and hired round-the-clock Pinkertons to guard his precious sperm. Baffled by the public rage, Graham retreated from the press. He ducked interviews for two years. His dream had become a farce.

Graham's other two Nobel donors quit when the bank went public; they feared a Shockley-style mauling. So by late 1980, Graham found himself presiding over a Nobel Prize sperm bank that had no Nobel Prize donors, no Nobel Prize sperm left in storage (it had all been shipped out), and no Nobel Prize babies. None of the first three women who'd been inseminated with Nobel sperm had gotten pregnant. In fact no one inseminated with the Nobel sperm ever got pregnant. The Nobel Prize sperm bank would never produce a single Nobel baby.

The bad publicity did not deter those who mattered most to Graham: customers. Every day the Nobel sperm bank was flayed in the press. But every day more applications—applications from desperate women—arrived at Graham's office.

And Graham remained an irrepressible evangelist for his cause.

He had customers but no Nobelists, so now he needed some new seed—some dazzling, backflipping, 175 IQ sperm—to give them. Graham was nothing if not a canny businessman. He had built a multimillion-dollar eyeglasses company that depended on a fickle public's taste. He understood that he had to please his customers. This was a radical notion in the fertility trade, where doctors were accustomed to bossing around their forlorn infertile patients. Graham had begun with the notion of limiting his clientele to women who belonged to Mensa or had an IQ of at least 120. "I don't want a whole flock of ordinary women," he had declared. Graham quickly ditched that plan. If he wanted a popular bank, he realized, he would have to take any woman—or at least any married woman with an infertile husband—who applied.

More important, Graham had learned that his customers didn't share his enthusiasm for brainiacs. The Nobelists had afflicted Graham with three problems he hadn't anticipated: first, there were too few of them to meet the demand; second, they were too old, which raised the risk of genetic abnormalities and cut their sperm counts (a key reason why their seed didn't get anyone pregnant); third, they were too eggheaded. Even the customers of the Nobel sperm bank sought more than just big brains from their donors. Sure, sometimes his applicants asked how smart a donor was. But they usually asked how good-looking he was. And they *always* asked how tall he was. Nobody, Graham saw, ever chose the "short sperm." Graham realized he could make a virtue of necessity. He could take advantage of his Nobel drought to shed what he called the bank's "little bald professor" reputation. Graham began to hunt for Renaissance men instead—donors who were younger, taller, and better-looking than the laureates. "Those Nobelists," he would say scornfully, "they could never win a basketball game."

Graham dispatched his newly hired deputy, a former sanitation engineer named Paul Smith, to find another crop of donors, men who could write equations all morning, ski all afternoon, and make love all night. Graham had known Smith for almost twenty years. Smith had thought about little else than genius sperm banking since 1960, when, as a nineteen-year-old Berkeley undergraduate, he had read an article by Hermann Muller. Smith believed with a convert's passion that genius sperm banking could change the world. Smith had written fan letters to Muller and struck up a great friendship with the old scientist. (Muller even tried to marry Smith off to his daughter.) Smith had accompanied Muller to the 1963 Pasadena meeting where Muller and Graham wrote the initial plans for the sperm bank. Graham and Muller had even invited Smith to serve on their first advisory board in the early 1960s. But after earning an engineering degree at Berkeley, Paul fled to England in 1965 to avoid having to serve in Vietnam. He spent fifteen years in Britain, working as a printer and road engineer, then returned to the United States right before the launch of the Nobel Prize bank in 1980. When Smith heard the first news reports about the Repository, he hopped into his car, drove across the country to Escondido, knocked on Graham's door, and volunteered for the cause. Graham appointed him manager of the bank, offered him a $26,000-a-year salary, gave him a dealer-fresh, banana yellow VW Beetle, and told him to go forth and multiply.

In every way Paul Smith was Graham's opposite. Graham was short, Paul was tall. Graham was a neat freak, Paul a slob. Graham was gracious, Paul was socially awkward. Graham's mind was orderly, Paul's was anarchic. With his little head on his gaunt, endless body, Smith looked a little bit like a spermatozoon himself. He had the face of the actor Ed Harris and the manner of an addled but brilliant professor. He spoke in a flat, creaking voice; long pauses separated his words. Sometimes he scarcely seemed aware of the world around him. He walked like a John Cleese character, all knees and flailing elbows. He cared little for his personal appearance. His hair flew away in odd, mad-looking wisps. He loved dogs and bred them; he was constantly coated with a mat of dog hair. Yet he had a sharp mind, a wickedly dry wit, and, most important for Graham, a messianic zeal for genius sperm banking that surpassed Graham's own.

Smith recruited for Graham as if his life depended on it. He haunted the campuses of Caltech and Berkeley—he preferred Caltech because "parking was easier." He pored over *Who's Who* and scanned lists of Fields Medalists and other hotshot scientists, seeking men who were brilliant, good-looking, and willing. His luck was spotty: he landed only eight donors in more than a hundred solicitations. "Some of the men had a vasectomy. Some said their wife wouldn't let them. And some probably thought I had been sent by the Devil himself." Smith would go to almost any length to get his sperm. On the rare occasions when someone agreed to give, Smith would immediately drag the donor somewhere—anywhere—where he could jack off. That might mean a quickie motel, or worse. One donor recalls Smith handing him a cup and directing him into a public bathroom at the University of California, Berkeley (surely not the first time a Berkeley bathroom was used for a curious sexual purpose, but still . . .). Smith's finds included a prize math professor, student prodigies, and dazzling computer scientists. Meanwhile Graham, working on his own, recruited others, such as Edward Burnham and the Olympic gold medal–winning Donor Fuchsia.

From the beginning, the Repository had more applicants than sperm—a state that would persist until it closed nineteen years later. Every sperm donation produced only a handful of usable semen straws. But each client needed a couple of straws every month, and it usually took several months to get pregnant. Graham couldn't sign up donors fast enough and couldn't get them to donate often enough to meet the demand. They were busy men, and Graham didn't pay them, so it was hard to persuade them to donate frequently. He and his employees ended up taking a triage approach, supplying what little sperm they had to the women who seemed most likely, or most desperate, to get pregnant.

Graham didn't enlist any minority donors, mostly but not *en-*

tirely for lack of trying. Graham's racial views were complicated. He retained the childhood prejudices learned in Harbor Springs. He felt uncomfortable around blacks and considered them mentally inferior to whites. Like Shockley, he felt his conclusion was scientific, not emotional. The "IQ of the average Negro," he would say with certainty, is "twenty-two points lower than that of the average white American's." (That was ten more points than even Shockley claimed.) When Lori Andrews, a lawyer and bioethicist, applied to Graham for Nobel sperm as a kind of stunt, he asked her for legal help: he said he wanted to find a way he could give sperm to her—a single white woman—but not to a single black woman.

Although Graham shared Shockley's racist conclusions, he didn't make a federal case of them. For Graham, race was a footnote to the larger story of genetic decline everywhere. He was *willing* to believe that genius could appear in anyone. But his fixation on IQ made him think it was much less likely among blacks and Hispanics. Graham was a racist but not always a white supremacist. He ranked blacks and Hispanics below whites in intelligence, but he ranked Asians above whites. He frequently tried, and always failed, to recruit Asian donors. And though he had grown up in a distinctly anti-Semitic town, Graham hugely admired Jews, whom he believed to be disproportionately intelligent. A large number of his donors were Jewish.

Graham's prejudice didn't stop him from trying to recruit blacks. When Smith—who didn't share Graham's racial views—identified a brilliant black donor candidate, Graham encouraged him. They were both disappointed when the man's diabetes disqualified him. The bank remained lily white. According to three former Repository managers, no black women ever applied for Repository sperm. This isn't too surprising, since there were no black donors and people usually choose a sperm donor of their own race. The Repository did have Asian and Arab applicants who got pregnant.

In the early 1980s, Graham worked hard to turn the Repository into a respectable business, rather than a ludicrous one: Graham's wife didn't like keeping the sperm at the Escondido estate. Not only had the house been picketed, but a Japanese trespasser had once made a run at the sperm, only to be nipped by a family dog. So Graham gave Paul Smith a room at his Del Mar beach house and moved the sperm tanks down there with him. Finally, in the mid-1980s, Graham resettled the bank in a three-room suite in downtown Escondido. The office was in the back of a real bank building, a source of amusement to visitors.

Graham donated generously to Republican causes, but his notoriety hurt him. Some politicians he supported—Governor Pete Wilson, for example—shied away from him at political events (though Graham did score an invitation to the Reagan White House, among a group of other California businessmen). His reputation didn't dampen the attendance at his famous parties. Graham would throw open "Shadybrook"—whose manicured ten-acre grounds were Graham's other pride and joy—serve milk he had milked from his own goats, show off his horses, and bring in an old-fashioned ice cream wagon to scoop Baskin-Robbins. Graham sometimes invited donors to the parties and would slyly drop hints to them that other donors were in the room.

Graham reveled in the theater of the genius sperm business. He grandly instructed his employees to "go out and change the world!" Whenever a reporter visited—which was often after Graham lifted his interview ban—Graham made a great show of removing his sport coat and pulling on a short white lab coat, as if to say: Here I am, a man of science. (In fact, nothing that qualified as science was practiced at the Repository. It was a mailroom, not a lab; its main activity was packing frozen vials into shipping canisters and handing them over to the Federal Express guy.) Still, Graham relished posing for photos in his lab coat. He liked photographers to shoot him as he opened a liquid-nitrogen vat, so that he was wreathed in a billowing cloud of vapor. The pictures were weird and compelling—a perfect, if ominous, advertisement for the product.

Even more than the PR, Graham adored the process of recruiting donors. To most people, Graham was polite but cool. With his donors—with any man he admired—he fawned. Except for his vanity about his looks, Graham wasn't very impressed with himself. Despite his wealth and extraordinary creativity, Graham never felt he belonged in the company of the men he was recruiting. He could easily have donated sperm to his own bank, but he never did so—he said that he was too short and not accomplished enough. It was an odd personality quirk. Even while conscripting donors who were manifestly less intelligent, less creative, and less good-looking than he was, Graham believed them worthier of posterity. Graham gushed about them: "These are just dandy guys. They have it all. They have the physique, the looks, the brains, that it would be good for everybody to have. There wouldn't be a lot of political problems and economic problems in the world if everyone was like these guys."

Graham seduced his targets like a lover. He literally got crushes on them. (He wanted to have their babies.) He arranged meetings as if they were first dates. He dressed to the nines—immaculate sport jacket, tie, pressed white shirt, black leather belt with a silver "RKG" buckle. He took them to long dinners—he was a famously slow eater—where he flattered them with questions about their work, their life, their families. He studied up on his men—not merely the Nobelists, but even later donors whose achievements were meager—reading their academic papers, memorizing details from their entry in *Who's Who of Emerging Leaders*. He didn't use first names; anyone who could conceivably be called "Doctor," he called "Doctor."

If he was traveling to see a donor—and he was known to fly across the country for a quick meeting—he would invite the donor back to his hotel room after dinner, to try to collect a first sample. He would work the candidates over to lure them to his room. "The oddest thing about contributing was that I felt like a girl. I felt feminine," says one of Graham's donors. "He took me to dinner, he tried to get me back to his hotel room. . . . And I said, 'Now I know what women feel like.' " The recruits often felt powerless to resist Graham. He was so flattering, and the donations mattered so much to him. They couldn't bear to disappoint him. Graham didn't pay his donors—he was the only sperm banker who didn't pay—but that was part of the seduction. He made them *want* to do it.

And when he found his man, how thrilled he was! He took one look at Donor Orange-Red and exclaimed, "What a specimen!" The first time Graham persuaded Edward Burnham to give a sperm sample, Graham paced outside the bathroom anxiously, like an expectant father. When Burnham emerged with the cup, Graham grabbed it from his hand, swabbed some of the semen on a slide, rushed over to the microscope, and peered down at Burnham's spunk. After a moment, he stood up, made a fist, and shouted, "Yes! You're just the man I thought you were!"

By 1982, Graham had good donors; he had sperm; he had customers. But he still didn't have babies. Twenty-five years ago, artificial insemination with frozen sperm was a crapshoot. (Today's ovulation kits and hormone shots have boosted success rates considerably.) Finally, in April 1982, the hex was broken. An Arizona couple named Joyce and Jack Kowalski gave birth to the Nobel Prize sperm bank's first baby, Victoria Kowalski. Her birth remained secret till June 29, when *The National Enquirer* broke the story with a two-page spread: "Mother of First 'Nobel' Sperm Bank Baby Tells Her Incredible Story: Our Miracle Baby Could Be America's Hope for the Future." Months earlier, the *Enquirer* had approached Graham's assistant Paul Smith, offering $20,000 for an exclusive on the first baby. He had passed the *Enquirer* on to the expecting Kowalskis.

The *Enquirer* feature, in classic tabloid fashion, revolted and mesmerized at the same time. Joyce was said to have chosen Victoria's donor father from a list of "the greatest minds of our time,"

picking "a famed mathematician with a genius IQ of over 200."
Joyce described Victoria as "a baby who could be the first of a new
breed of genius children." She recalled cradling Victoria in her
arms for the first time. "Her dark blue eyes flashed with intelligence. 'What will she become?' I mused. 'A female Thomas Edison
or Einstein.' . . . I imagine her as a young child studying college
textbooks. I see her working out complex mathematical equations
quicker than a computer."

Whatever dim views readers had of a genius-sperm-bank-going
parent, Jack Kowalski confirmed them. He told the *National Enquirer*, "We feel we have a duty to raise this child for the betterment
of society. We'll give her the best educational opportunities possible.
We'll begin training Victoria on computers when she's 3, and we'll
teach her words and numbers before she can walk."

Instead of restoring the luster Shockley had stolen from the
Repository, the Kowalskis tarnished it again. Graham and Smith
hadn't vetted them carefully enough. As reporters chasing the *Enquirer* scoop soon discovered, the Kowalskis were convicted felons.
When they applied to the bank, the couple had just been sprung
from federal prison, where they had served a year for using the identities of dead children to secure credit cards and bank loans. That
wasn't the worst of it: the Kowalskis had also lost custody of Joyce's
two children by her first husband. According to the first husband,
Jack Kowalski had sent his stepson to school in his pajamas wearing
a sign that said he was a bed wetter. Jack Kowalski also apparently
tried to bully the first two kids into becoming prodigies, abusing
them if they didn't work hard enough: "My daughter was made to
wash dishes in superhot water if she did not toe the academic line,"
Joyce's first husband alleged. The Kowalskis disappeared after the
press started pursuing them, eventually settling in backwoods
Arkansas and homeschooling Victoria.

The Kowalski fiasco reflected the Nobel bank's general chaos.
Graham was a hands-off manager, and his employees ran the bank

haphazardly. Records were a mess, when they were kept at all. Often, the bank didn't track what sperm it gave to what mother. The donor-coding system was inconsistent. Some donors were numbers; others were colors; others were numbers with colors. Sometimes colors were reused: there were at least two different Donor Oranges, and a Donor Orange/Red. There was a Donor Yellow, a Donor Brown, and a Donor Yellow/Brown. The donor catalogs were rife with misspellings. It couldn't have made a great first impression when applicants read about "Donor Corral," "Donor Fucshia," and "Donor Turquois." Some genius!

Paul Smith's eccentricities were also troubling, at least to his assistant, Julianna McKillop. A disgruntled customer sued the bank, claiming that the vials sent by Paul had contained barely any sperm, certainly not enough to get her pregnant. Graham settled the lawsuit out of court. A little later, Smith drew the bank another suit, this one for libel. He told an interviewer that any woman who wanted to bear "defectives" should go to the Sperm Bank of Northern California. The rival bank sued. Smith was something of a showman, and his stunts were not usually helpful to the Repository. To ensure his anonymity—and thus protect the identities of his donors—Smith would be photographed only when wearing a surgical mask. He would even do TV interviews wearing the mask—an affectation that seemed less a useful precaution than an oddity.

But never mind, women kept applying. The *National Enquirer* story produced another hundred applicants. A bunch of pleading letters followed every media mention. Nothing helped the bank's reputation more than little Doron Blake, the bank's second baby, born in August 1982. His mother, Afton Blake, had failed to get pregnant by her first choice, one of the Nobel laureates. She had managed to conceive by Donor Red #28—a computer scientist and classical musician. (He also suffered "slightly" from hemorrhoids, according to his catalog entry.)

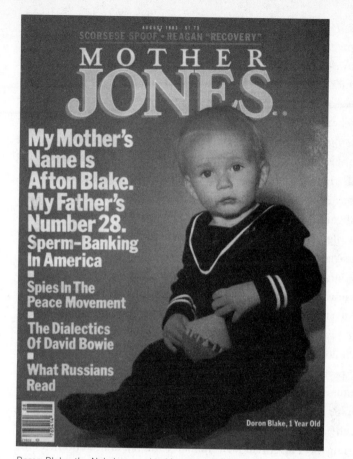

AUGUST 1983 $1 75

SCORSESE SPOOF · REAGAN "RECOVERY"

MOTHER
JONES..

**My Mother's
Name Is
Afton Blake.
My Father's
Number 28.
Sperm-Banking
In America**

■

**Spies In The
Peace Movement**

■

**The Dialectics
Of David Bowie**

■

**What Russians
Read**

Doron Blake, 1 Year Old

Doron Blake, the Nobel sperm bank's celebrity baby, on the cover of the August 1983 *Mother Jones. Courtesy of* Mother Jones

Afton was a "transpersonal" psychologist; one of her specialties was plumbing her patients' past lives. She was a shy, socially clumsy woman, but she had the psychologist's faith in openness. Secrets were poison; everyone should tell everything. (Within minutes of meeting me, for example, Afton was sharing her sexual dreams about Bill Clinton.) The media appetite for superbabies was gluttonous, and Afton was happy to feed it. She did interviews while she was pregnant. She went on TV with Doron when he was two weeks old. At two months she posed him for *People*, a straw-haired little charmer. She put him on the cover of *Mother Jones* at a year, in *Maclean's* a couple of years later. She let *California* magazine ride the school bus with him, showed him off on *48 Hours*, *60 Minutes*, *Primetime Live*, you name it. *The New York Times* even published an editorial for the occasion of Doron's first birthday, "An Ideal Husband." It wrote of Afton, "She knows nothing but good of [the donor]. Her son can have nothing but pride in his amusing, athletic, intelligent, music-loving Dad. And Dad, who was to Doron's mother what the bee is to the flower, was equally free to buzz away. Why ruin so perfect a relationship?"

Doron obliged by being a prodigy, or so his mother kept reminding everyone. "Doron" was Greek for "gift," as Afton would tell reporters. This produced stories that began by marveling at the many "gifts" of the boy named "Gift." Afton let a reporter arrange a psychological exam for Doron when he was all of four months old. The psychologist declared that Doron had a mental age of eight months. As a newborn, Doron could mark time to classical music with his hands. By age two, he was using a computer. He was playing chess at five. By kindergarten, he was scanning the *Iliad* and learning algebra. A year later, he had written a draft of his first children's book. Two years later, he was poring over Einstein's theory of relativity.

All of this was passed on to reporters by Afton, along with even more intimate information: that Afton had breast-fed him till he was six. That she had bought him a subscription to *Playboy* so he could learn about sexuality. That she wanted him to grow up to be like Gandhi or Churchill.

Here was what Afton did not tell reporters but what they witnessed (and reported) anyway: Doron boasting of his 180 IQ;

Doron's classmates mocking him as "Sperm Boy"; Doron bragging, bragging—indulged and egged on by his mother and grandmother.

> INTERVIEWER: You read *Hamlet* in kindergarten?
> LITTLE DORON: Good gosh. Can't everybody?

Graham was thrilled with Doron. He sent the little boy books and drove up to Los Angeles to take the Blakes to dinner. Whenever anyone doubted the Nobel sperm bank, he countered by reciting Doron's accomplishments.

Doron sent Graham a photo inscribed, "Thank you for bringing me into the world." The photo took place of honor in the growing collage on Graham's office wall. By the mid-1980s, the bank was spawning a dozen kids a year. Graham mounted their baby photos in the Repository office. (Almost all were blond—which didn't help Graham shake the Nazi complaint.) Graham planned follow-up research to prove the kids' brilliance. In the meantime, he relied on anecdotes about Doron and unsupported assertions, confidently telling interviewers—based on no data—that most of the kids were well above average.

Doron Blake became Graham's mascot. Women read about the prodigy boy or saw him on the television, and they responded—sometimes with delight, sometimes with horror, always with fascination. By 1984, two years after Doron's birth, more than one thousand women had applied for Graham's special sperm.

CHAPTER 7

A FAMILY OF BASTARDS

After Samantha Grant told me she had found Dr. Jeremy Taft, the Miami plastic surgeon she was sure was Donor Coral, I avoided Tom for a few months. Not calling Tom removed any temptation I might have to spill the secret to him. I also wanted to give him time to get to know his half brother Alton, Samantha's son.

A few nights before Christmas 2002—eighteen months after we first talked—Tom phoned me at home. After a couple of pleasantries, he dropped the bomb: "I'm gonna be a father—a boy, a son. On January 31, that's when he's due. We're going to call him Darian Jacob Legare."

I was so addled that the only thing I could think to say was "January 31, that's my birthday, too." I did some quick math in my head and figured out that Tom was all of seventeen years old. The last time we had spoken, he had seemed barely old enough to think about girls, much less get them pregnant. I recovered enough to congratulate him and ask who the mother was. (I hadn't yet seen the e-mails he had sent to Alton mentioning a girlfriend.)

"Her name's Lana. She's Russian. And I forgot to say, we're getting married. I met her a year ago—she was in my class at school. I'm in love. I really am. No one believes me, but I actually proposed to her before I knew about the baby. I proposed on our six-month anniversary, and a few days later she found out she was pregnant."

I asked, doubtfully, how they were going to support themselves. Tom said he had a plan, sort of. He had graduated from high school a year early, and he was already more than halfway to his associate's degree. Lana would stay home with the baby for a while, and then she'd get certified as a massage therapist. Oh, and since she and her parents were illegal immigrants, Tom would figure out how to get her a green card once they got married. As for his own career, he was working at his mom's company. The job wasn't great—he was a file clerk—but they liked him and gave him full benefits. The company also paid for his classes, which he took nights at the local community college. He and Lana didn't have their own place, so they would split time between her parents' house and his mom's. By the time Tom finished explaining all this, he sounded exhausted—and overwhelmed.

The conversation was depressing me. Building a life out of string, chewing gum, and paper clips was hard enough for anyone, but when you were seventeen and engaged to a pregnant illegal immigrant . . .

Tom kept talking. He said he had called because he had a question for me, something he and his mom had been wondering about. "Am I going to be the first kid from the sperm bank with my own child? Is this the first grandchild from the Nobel bank?" There were only a couple dozen Repository kids older than Tom, and I hadn't heard about any who were married or parents. I told him that I didn't know for sure but that I would be shocked if he wasn't the first parent. "All right!" he exclaimed. "I set the record!"

Tom and I had known each other for a year and a half, but never met, so Tom invited me to visit him in Kansas City after Darian's birth. Tom was excited about my visit. Being a Nobel sperm baby was what made him special. Until I had come along with my odd questions, no one had ever taken an extraordinary interest in him; he had been a regular kid from a regular dysfunctional family living in a regular suburb of a regular city. He had never done anything remarkable—never broken a school record or won a state prize. For the first time, he was at the center of something that the outside world cared about. He was flattered that I would travel all the way from Washington just to meet him. I also sensed that Tom wanted to see me because he was feeling daunted by impending fatherhood. He didn't know a lot of fathers—his friends were all seventeen, after all. I had talked to him a little about what it was like to be a dad. I didn't think Tom desired my advice, exactly, but maybe he welcomed the idea of some father-bonding with me.

I flew out to Kansas City on a Friday morning in March, a couple months after Darian's birth. The day was raw and gray. I drove through endless suburbs. The Legares lived in practically the last subdivision outside the city; just beyond their development stretched fields, all the way to Saint Louis, perhaps. Their house had beige siding and green shutters. It looked friendly. There were cute little statues of squirrels on the lawn.

It was midafternoon by the time I finally rang the doorbell. A middle-aged woman answered the door and introduced herself as Mary, Tom's mother. Tom was still at work, she said. She added that her husband, Alvin—the way she said "husband" made it clear that the word had only a technical meaning for her—was on the road, as usual. Mary introduced me to Tom's fiancée, Lana, who greeted me halfheartedly. She was sprawled on the living room couch watching a monstrous TV. Lana was small and pale, with a pretty Slavic face and owlish glasses. She was wearing black nail polish and a creepy black Insane Clown Posse T-shirt. Every few minutes she would check on baby Darian, who was dozing serenely in a swing next to her. He was blond, square-headed, and cute.

There was very little furniture in the house, no family pictures, no sense that it belonged to anyone. I was surprised, because Mary had seemed so familial. As soon as she hung up my coat, Mary gestured to the empty spaces and said that it was not her house, it was a temporary rental. Her house, which was a few blocks away, had burned down the month before, she said. They had lost almost everything, including a dog and two cats. That was why they had so few things.

Mary and I sat down in the kitchen to talk before Tom got home. She was short, red-haired, pretty, and plump—though down forty-six pounds on the Atkins diet, thank you very much. Her eyes were bright and blue; she was cheerful; and she was argumentative. Mary worked her tech support job from home, which allowed her to keep an eye on the comings and goings of Tom, Lana, and her daughter, Jessica. Mary said she liked having Tom and his friends around, eating her food and commandeering her TV. Being here meant that they were not out doing much worse. Just as she was telling me this, a friend of Tom's wandered into the house, helped himself to soda from the fridge, and flopped down on the couch next to Lana. He introduced himself as one of Tom's bandmates. He said he "kind of" lived there, or at least crashed on the floor a lot.

Mary and I were just out of Lana's earshot. Mary told me how worried she was about Tom and Lana. "I want them to get married now, as soon as possible," Mary said. "Lana could take the baby right back to Russia, and there is nothing we could do about it, unless they're married." Mary's suspicions ran deep: she said she had made Tom get a DNA test when Darian was born, just to make sure that Darian was his. He was. (Mary's DNA-test demand had been a bitter pill for Tom, he later told me. He was searching for his own father. Yet at the same time his mom was trying to get him to lose his own son.)

Mary had to field a call from work, so she pulled Jessica out of her bedroom and told her to talk to me. Jessica, who was fourteen, seemed as though she'd rather do anything else, but slouchily obliged. She barely resembled her mother or the pictures I had seen of Tom. She was very thin. Her face couldn't decide whether it was going to be beautiful or belligerent. Right then, it was both. She had deep-blue cat eyes, wide cheeks, and an expression that was all at once scornful, cynical, sullen, and smart. Jessica was the biological daughter of Donor Fuchsia, the Olympic gold medalist, which she had learned when Tom spilled the secret to her.

To make conversation, I asked her about the pentagram pendant hanging from the black choker around her neck. "I'm a Wiccan. We believe in hurting none." She had a languid, worldly wise voice. She showed me the book she'd been reading when Mary had interrupted her, *Witch Child*. I asked her about school. High school ruled over middle school, she said. She was a freshman, she loved it, her grades were dropping from As to Bs and Cs, and she didn't care.

Mary, who was now eavesdropping, dropped a manila folder on the table. Inside were letters from the Repository and a copy of the donor catalog. The pages were lightly browned at the edges.

"During the fire, I was going to go back in to get jewelry and valuables," Mary said. "I asked Jessica what she wanted, and she told me, 'I want you to get my dad's folder.' So I went and got it. It still smells like smoke." I sniffed it: it did smell singey. "It is very interesting that she wanted that. She was afraid that she would lose what little she did know about the donor."

Jessica leafed through the folder.

"That is the only thing that I have connecting me to my real dad," Jessica said. "That was the first thing that popped into my head. Even if it is a little thing, a piece of paper. I don't want to lose it."

Jessica kept talking about Donor Fuchsia and the sperm bank. Discovering her Nobel sperm origins, she said, had made her more tolerant, more decent. "I used to be much more of a bitch. Finding out made me look outside my box and realize there is a whole world out there. I have a dad out there, a real dad among those millions and millions of people. It made me stop judging and made me realize different kinds of people are normal. I am giving people a chance that I wasn't giving them. I realized my brother is not a long-haired freak." Before she had known the secret, she said, she and Tom had never talked. Now they were tight. She listened to his music, even dated one of his friends.

But unlike her brother, whose obsession was pulsatingly obvious, Jessica seemed blasé about her sperm origins. "I did have a lot of curiosity about the donor when I first found out. I wondered if I would ever get to meet him. But I don't need that to know who I am. Now it doesn't really matter to me."

I told her that I was acquainted with some of her half siblings—other kids born from Donor Fuchsia. Did she want to e-mail or meet them? She shrugged. She didn't care.

When I had talked to Mary about her daughter in earlier phone conversations, she had seemed to blame Jessica's falling grades, sudden lack of interest in ballet, and stoner friends on her new uncertainty about her identity. Learning about Fuchsia, Mary suggested, had rattled Jessica. Talking to Jessica, I wasn't sure about that. Maybe she wasn't slacking because she was confused about Donor Fuchsia. Maybe she was slacking because she was a teenage girl, and that is a teenage girl's job.

Tom arrived home from work at 5 P.M. My first impression was that Jessica had been right: he *was* a long-haired freak. He had a broad face, tending to plumpness. He had blue eyes beneath a deep brow, a wide, strong chin with a great cleft struck out of it. Blond stubble traced his jaw. He was tall, with broad shoulders, but he was carrying an extra twenty-five pounds in the belly. But his hair! It was dirty blond, straight, and thick. When he shook it out of its workday ponytail, it fell in a cascade halfway down his back. It was amazing hair, a lifetime of hair. Tom looked like a Goth—I don't mean a Goth kid, I mean an actual Goth. He would not have looked out of place wearing a leather jerkin and swinging a mace. (This is intended as a high compliment to Tom, an avid fantasy gamer.)

As soon as he greeted me, Tom rushed into his bedroom, stripped off his button-down work shirt, and slipped into a T-shirt for one of his favorite bands, Twiztid. The T-shirt said "Freek Show" and had a picture of some nutcase in white pancake makeup. Three quarters of Tom's wardrobe, I would discover, consisted of black T-shirts for Twiztid or Insane Clown Posse, all of them with pictures of nutcases in white pancake makeup.

Tom was accompanied by a bunch of friends. I couldn't really tell how many; sometimes there seemed to be three, sometimes four, sometimes six. Most of them were wearing band T-shirts and black trench coats. They all had long hair, and they all seemed to be named Mike or Matt (except the girl). They looked very Columbine, but in a sweet way. They were soft-spoken, polite, and good-natured—nerdy, not scary. Tom explained that his house had become a kind of clubhouse for his posse because he had the best video game setup, because his mom didn't mind having everyone around, and because he and Lana had to be home with the baby anyway. So every Friday night, and lots of other nights, a dozen of his friends came over and spent the night gaming and otherwise fooling around.

Mary shooed the Trench Coat Mafia out of the kitchen so we could talk. They headed to the basement and stayed up till 5 A.M. playing video games and three-dimensional chess, eating Twix bars, and—when more girls arrived—exchanging back rubs.

Tom, Mary, and I remained upstairs in the kitchen. Tom grabbed a full two-liter bottle of Sam's Cola from the fridge and started chugging straight from the bottle. By night's end, he had finished that whole bottle and a second one—filling him with enough sugar and caffeine to wake a dead man. In Tom's food pyramid, Sam's Cola ranked as the most important of the three basic food groups. Candy and Burger King were the other two.

Tom was glad to see me. Beneath an ominous appearance and mumbling voice, he possessed a profound sweetness and eagerness to please. Words tumbled out of him as he caught me up on his life. He was proud of Darian, proud of Lana, proud of his job, proud of his college classes—which he was managing to ace despite work and baby. He was thrilled to be engaged to Lana. He delightedly waved his hand in front of me to show off the promise ring Lana gave him and asked Lana to come over to show me hers. Tom's showing off was endearing, not cocky. He was doing the right things, even though they were hard, and he wanted to share that with me.

After a little bit, Tom badgered me to come downstairs so he could show off his video game setup. "Did you see the basement?" he asked me. "It's like we robbed an electronics store!" He led me down into the expansive playroom, where the Trench Coat Mafia was silently arrayed around one of the three computers, murdering aliens (or maybe one another) on Halo. "We have everything," Tom bragged. He pulled out his Sony PlayStation 2, an old Atari, an Xbox, a Sega Genesis, various Nintendo systems, and several game consoles I had never even heard of. Tom said he was even building an "arcade game emulation": he was going to mount a computer like a 1980s video arcade terminal and use it to play cool old-school games such as Space Invaders and Centipede. "I'm lucky I have Lana and Darian because otherwise I would spend all my paycheck on games," he said, not kidding at all. The basement also had a pool table, an air hockey table, and an army of Dungeons and Dragons metal figurines. He bumped one of his friends off Halo so that I could try it; much mirth ensued at my ineptitude. Tom and his friends traded insults in some weird gamer dialect. Someone said, "I put it on mimic, dude." Everyone laughed.

I asked Tom about his band, Infernal. He said that Infernal was dead but he had a new band, Durga. He was writing all the lyrics. Durga sounded like ICP and Twiztid, he said. A friend with a small studio had invited Durga to record a CD and was going to release it. Tom offered to play a track for me. He booted up another com-puter—the one he recorded on at home—and opened the Durga folder. He considered playing "I Wanna Fuck You" or "Ghost Cat" but clicked on "Shut My Eyes." Behind a thumping bass line, I heard Tom's voice:

No weapon is mightier than the pen.
The whole world goes dead when I shut my eyes.

It sounded awful to me—a horrible combination of screaming and droning. But I could hear my own adolescence in it. If I were seventeen again, I would probably like it a lot. It was relentless, loud, and passionate—a close cousin to the barbarous testosterone-laden crap that I listened to at Tom's age.

Upstairs, Darian had awakened. As soon as we heard him, Tom headed back up to the living room. Tom hoisted him out of his baby swing and carried him over to me, displaying him like a trophy. Tom beamed at Lana, the boy, and me. "The first grandchild from the Repository!" he announced. Darian whimpered. "I think he needs a bottle," Tom said. He trotted to the fridge, poured some sterilized water, measured out the formula, shook the bottle up, and started feeding it to his son. It was a sweet and incongruous moment—a seventeen-year-old boy in a sociopathic T-shirt who would rather be playing Halo with the Columbine Crew, patiently feeding and burping his son.

"I don't really feel like a dad a lot of the time. It's like I am babysitting. I am not responsible enough yet," Tom said. "I get yelled at for playing too many video games, for not waking up to feed him. He comes first now. That's really new to me."

Lana smiled and interrupted Tom. "You're a very good dad. You have a hard time waking up to take care of him, but you are a very good dad."

As Darian finished the bottle, a guy and a girl emerged from the basement and ducked into Tom's bedroom. Tom quickly excused himself, walked over to the bedroom, and beckoned them to come out. A little shamefaced, the couple returned to the basement. Tom came back, annoyed and amused. Whenever those two were left alone, Tom explained, they sneaked off to have sex. They shouldn't be doing it, he said, and they definitely shouldn't be doing it in his room without his permission.

I suddenly realized how condescending I had been toward Tom. I had been smugly thinking of him as the long-haired oddball who had knocked up his illegal immigrant girlfriend, who was scraping along with a dead-end job and only the vaguest idea of a future.

I was all wrong. Tom had grown up in a dreary suburb where kids dreamed small. Tom was the only one of his friends who held a real job (almost the only one who held any job). He was taking a full load of classes; his friends barely managed one. He was the guy who was writing the songs and cutting an album; it might never sell more than seventy copies, but he was doing it. He had gotten Lana pregnant, but he was doing right by her. He was marrying her, raising their son, getting her a green card, supporting her through school. Yes, he lived in his mom's house. Yes, he wanted to goof around and play Halo all the time. He was still a teenage boy, after all. But he was seventeen going on forty-three. He wasn't a genius. He knew that. But he was capable, and that counted for a lot more in his life. He was keeping it together when no one else was bothering to.

The family piled into my rental car for a ride to a nearby restaurant. I asked Tom how he and Alton were getting along, what it was like to have a brother. Tom slumped in the front seat. He said they had stopped e-mailing a while ago. They had had "trouble relating," and Samantha had suggested to Mary that the brothers take a breather, give Alton a little more time to grow up.

"His mom said I was a bit too 'mature' for Alton. I had told him I was in group therapy and why I was in it, and I think she didn't like that. She said maybe we should get back in touch in a year." Tom paused for a second, rankled by the "mature" line. "However much Alton is maturing, I am having to mature *a lot* more." He nodded toward Lana and Darian. Tom tried to play it cool about the breakup with Alton, but he couldn't hide his disappointment. Again and again during my visit, he would return to Alton and Samantha's rejection. Tom was baffled by their brush-off. His brother had been yanked away, and no one would tell him why.

To change the subject, I asked Tom about why he had chosen "Darian" as the pseudonym for his son. Tom adored fantasy novels, and I knew that his son's real name had come from one of his favorite fantasy novels. The pseudonym had, too, Tom said. Darian was the hero of a book called *Owlflight*. "It's about a boy who loses his parents and then finds himself, finds love, and ultimately finds them again in the end." Tom's symbolism was not lost on me. He had found love in the form of Lana; now he needed to find the lost parent, the vanished Donor Coral.

Tom started talking about "him." He didn't have to say who that was. "It is just not fair. I don't even know his name, and I don't think I will ever know it. For Darian, I wish I could find out who his grandpa is. And for me, I *really* want to know. I want to know who he is and what he did. And I would like to meet him, even just once."

Tom asked me if I knew where the Repository's records were stored. I said that I wasn't sure but that Hazel in San Diego might have them. Tom thought about this for a moment, then said, "I'm thinking of going out to California and getting a job with whoever has the records—not tell them who I am—and then sneaking a look at my file." I started to laugh at this but realized Tom was serious.

I had done enough reading about children of donor insemination to know that Tom was unusual. Most "DI" kids his age aren't interested in their donor fathers. According to psychologists I have talked to, DI kids tend to be most curious about their donor fathers just *before* adolescence. That's when kids start to construct their own identities, when they are still attached to their parents but breaking away. It's at that age when kids fantasize that their parents are pirates or princes, so it's understandable that a lost dad would fascinate them. But once adolescence hits, the interest usually wanes; they form friendships that matter more than family, fall in love, make their own way in the world. Only much later, when they are married and having their own children, do they wonder again about their genetic origins and lost dads. Tom was in the heart of adolescence, yet he was fascinated with his donor father. I told him how exceptional that was, that he was acting more like a thirty-two-year-old than a seventeen-year-old.

Tom puzzled over it, then said, "But it makes sense for me, because I was becoming a dad myself. I guess that's why I have been thinking about him so much."

He continued, "When we found out Lana was pregnant, we signed up for this welfare program to pay for pregnancy and birth. They sent someone over to the house to tell us about pregnancy. The person usually gave Lana some homework. And one time she gave Lana a family tree to fill out. Lana was supposed to fill it out for her side of the family, and I was supposed to fill it out for my side. But it was pretty awful when I sat down to do it. My mom does not know who her real dad is. My dad—that is, her husband Alvin, not my donor dad—doesn't know who *his* dad is. And then my dad is not even my dad, and I don't know who my dad is. Basically, we are a family of bastards."

"So what did you write down on the family tree?" I asked.

"I just put question marks. It was depressing."

The man Tom had considered his father for seventeen years was no longer really his father. The man who was Tom's biological father was missing. The boy who had briefly been Tom's brother was no longer talking to him. Tom's life was full of absence.

Tom blamed the emptiness on his mother, mostly. Her mistakes had left his life in such a mess. Physically, Tom and Mary were not much alike, but their emotional temperature was similar. Mary was abrupt and Tom was naive, but they had the same spirit of openness. They told each other the truth, or a lot more of it than mother and son usually did. Lana's arrival had divided them a little—Tom was torn between his two women. And Tom was growing up. Their relationship was the heart of the family, and it was fraying. His search for his donor father had placed new strains on it.

Tom's anger at Mary poured out at dinner. Tom started by razzing her for having chosen the Repository. He said she hadn't considered how going to a genius sperm bank would mess up the family.

"Because I went there, I think you have advantages that other people don't have," Mary countered. She turned to me and made her case: "What I did gave them a better chance in life. I know things were always easy for my kids, maybe because they had better genes. I did not have problems that other parents did. Yes, I did focus on their grades. I had a fit when Jessica had a D, and I think I should have. Not enough parents do."

Tom responded. "The bank was selfish for you and the donor. The donor said, 'I am better than the average person, so I should be in this sperm bank.' In theory, it looked like a good idea, but when you get down to it, it is a Nazi idea."

Mary got more defensive. "Don't you think that girls always seek certain qualities in men? What is the difference if you do it with a donor or a boyfriend?"

"You never got to know the donor," said Tom. "You were presented with a sheet of paper. You could only make the choices they gave you to make."

Mary answered, "You don't understand. I got married at nineteen. I was married for six years and kept trying to have a baby and I couldn't. You don't know what that is like." Mary looked pointedly at Darian and Lana, then went on, "I still think the Repository was good, and I don't like it when you suggest I did something wrong."

"I *don't* think you did something wrong. I think you made the best of a bad situation. But think about what it's like for me. I *can't* know who my dad is," said Tom.

Tom started to complain about Alvin, his dad, her husband. "I remember that he would call us 'your kids.' He used to tell you, 'Clean up after your kids.' "

I asked Jessica if Alvin was now aware that she knew the truth about him. Jessica smirked. "I think Mom just told my dad by accident." Mary looked pained. Mary said that when she had spoken to Alvin on the phone the day before, she had mentioned that I was coming to talk to Jessica and Tom. "He said, 'Oh, Jessica knows now, too?' Then he was really quiet."

None of the Legares had much sympathy for Alvin—they knew him too well, I guess. Still, I couldn't imagine how humiliating this must have been for him. Maybe he had been an indifferent father to Jessica, but at least he had been her father. Now she knew he wasn't even that.

Mary took this opening to mention that she was planning to serve divorce papers on Alvin when he returned home from his current road trip. Tom had been expecting a divorce for a long time. Now that it had arrived, it was bittersweet. He was glad for his mom, who was certainly going to be happier. But he felt a little sad for himself, too. He had lost his dad once when he had learned about the Repository; now he was losing him again to divorce.

After Mary mentioned the divorce, I told the Legares about one of the odd things I had noticed in my reporting on the genius sperm bank: in most of the two dozen families I had dealt with, the father was notably absent from family life. I knew I had a skewed sample: divorced mothers tended to contact me because they were more open about their secret—not needing to protect the father anymore—and because they were seeking new relatives for their kids. I had heard from only a couple of intact families with attentive dads. While good studies on DI families don't seem to exist (at least I have not found them), anecdotes about them suggest that there is frequently a gap between fathers and their putative children. "Social fathers"—the industry term for the nonbiological dads—have it tough, I told the Legares. They are drained by having to pretend that children are theirs when they aren't; it takes a good actor and an extraordinary man to overlook the fact that his wife has picked another man to father his child. It's no wonder that the paternal bond can be hard to maintain. When a couple *adopts* a child, both parents share a genetic distance from the kid. But in DI families, the relationships tend to be asymmetric: the genetically connected mothers are close to their kids, the unconnected fathers are distant. I suspected that the Nobel sperm bank had exaggerated this asymmetry, since donors had been chosen because mothers thought they were *better* than their husbands—Nobelists, Olympians, men at the top of their field, men with no health blemishes, with good looks, with high IQs. Of course sterile, disappointed husbands would have a hard time competing with all that.

Robert Graham had miscalculated human nature. He had assumed that sterile husbands would be *eager* to have their wives impregnated with great sperm donors, that they would think more about their children than their own egos. But they weren't all eager, of course. How could they have been eager? Some were angry at themselves (for their infertility), their wives (for seeking a genius sperm donor), and their kids (for being not quite their kids). Graham had limited his genius sperm to married couples in the belief that such families would be stronger, because the husbands would be so supportive. In fact, Graham's brilliant sperm may have had the opposite effect; I told the Legares about a mom I knew who said the Repository had broken up her marriage. Her husband had felt as though he couldn't compete with the donor and had walked out.

When I finished, Mary responded that the sperm bank hadn't shattered her marriage. It had been doomed anyway. But Tom said he thought the rest of my description of strained families applied to them. He and Jessica were very alienated from their dad, and maybe that was the sperm bank's fault. Tom angrily challenged his mom: "Why didn't you stop to think about the gap we would have with Dad?"

"I didn't think about it because I did not know about it," Mary answered. "No one knew about that. I thought we were just going to be a family. In the early days, Dad took full credit. He thought you were cute."

I defended Mary, too. Seventeen years ago, no one was particularly worried that fathers would reject their own nonbiological kids.

Mary spoke more gently to Tom. She said she was sorry Alvin had not turned out to be a particularly attentive dad to him, but that the sperm donor wasn't necessarily the ideal dad, either. "You are looking for the father you wanted to have but didn't have."

Tom nodded and looked resigned. "Yeah, and even if I find Donor Coral he probably wouldn't be that father."

"I don't want you guys feeling you should be ashamed of your origins," Mary said.

"I am not. I am not," said Tom. "Look, I am very happy you did what you did. You don't have to be defensive about it. It's great that you found a good sperm donor. Honest, I don't think Dad would have been a good dad anyway, no matter what."

Mary tried to sum it all up optimistically. "I think you should look on the sperm bank as a positive. You can go through school without studying for tests." That was our last word on the subject. We retreated to the calmer, duller conversational topics of family life: Darian's sleeping habits, Tom's classes, Mary's job.

But the dinner argument stuck with me. All parents expect too much of their children. The United States is beset by tennis parents, aggro soccer dads, and homeschooling enthusiasts plotting their children's future one spelling bee at a time. Mary wasn't that bad. But she certainly did goad her kids to do better, and definitely hoped that knowing about their special origins would inspire them.

But in the case of a sensitive soul like Tom, I wondered if genetic expectation had inspired him or punished him. When your mom tells you you have to do better, you try to do better. But when your mom tells you your genes say you have to do better, it's different. You lose your free will. In some ways, the only logical response is to rebel and screw up, just to prove that your genes don't rule you.

Tom was too dutiful a son and too responsible a father to rebel on purpose, but when I left the Legares the next day, I got the sense that he was starting to feel taxed by his genes. He had been given a fantasy of Donor Coral—Dad was brilliant, handsome, healthy, and kind. But the fantasy was also a burden. How could he possibly live up to it? He was expected to find the magical, mysterious Coral-planted genius locked in his DNA and do something extraordinary with it. But he was already raising a son, holding a job, earning a degree, marrying a girl. Wasn't that enough?

CHAPTER 8

THE SECRET OF DONOR WHITE

A year passed, and I didn't hear from Donor White. Lots of people offered to help me find him. Two private detectives volunteered to search. TV producers kept calling: if Beth and Joy appeared on their show, they said, it would definitely smoke out Donor White. I conveyed all the requests to Beth. At low moments she entertained the idea of going on television, but she eventually rejected the idea: finding Donor White would be nice, she said, but not if the price was sacrificing the family's privacy and Joy's innocence. A hope was all it had ever been. "If nothing comes of it, I will have lost nothing. I knew it was a long shot." Both of us gave up on Donor White.

I took a three-month work trip to Japan and forgot about sperm banks. A few days after I returned to America, on June 12, 2002, I logged in to my e-mail and saw a message waiting from "rwhite6@aol.com." It began like this:

> Dear David Plotz,
>   This is Donor White and, even though some 15 months
> late, I hope that you will be so kind as to pass on this note

and my e-mail address to Beth about whom you wrote in your article regarding the Repository of Germinal Choice (RGC).

The e-mail continued for 2,300 words. "Donor White" described how the Repository had recruited him. He recounted some of his family history. He sheepishly mocked his own Internet incompetence to explain why it had taken him so long to see the article:

I am sorry to be so late in responding, but some allowances should be made for lack of knowledge about the type of Internet search engines that finally led me to your article, considering that I was one of those who went to college in the days when students wore their foot-long slide rules dangling from their belts and tied to one leg like a gun fighter in the Old West. Later, when introduced to computers, I carried a foot-long tray of punched cards into a room about the size of a basketball court, all of which was required to hold a single computer. Those of my generation can never compete in cyberspace with younger people who grew up using modern computers.

The letter finished sweetly:

I cannot imagine that some of the donors contacted have said that they rarely think about their children, because I think of mine very often. Indeed, I expect that they will be included among my last conscious thoughts on this sweet earth.

My thanks and best regards,
Donor White

Donor White sounded like none of the other donors I had talked to. Until Donor White, the donors had split neatly into two categories: the rationalists and the egotists. The rationalists, such as Edward Burnham, were matter-of-fact. They summed up the experience of having donated to a genius sperm bank with a shrug. They weren't troubled by it, and they weren't delighted by it. They weren't really interested in it. They didn't care about their "children." For them, donating to Graham was a nearly forgotten favor.

The egotists—such as Michael the Nobelist's son—were obsessive, creepy, and self-aggrandizing. Donating to the bank had been the greatest moment of their lives—not because they had helped anyone but because they had hoodwinked Mother Nature. They cared about the children the way a miser cares about gold. The purpose of the children was to be counted.

But Donor White was different. In form, he resembled the rationalists. Donor White wrote his letter with the exactitude of a scientist—dates and times recalled precisely, names spelled right, all facts crisp. The prose was formal, even pedantic: ". . . about whom you wrote in your article regarding the Repository of Germinal Choice (RGC)." But the soul of the letter was something else, something new. His language, for example: it was long-winded, but it was courtly. (I immediately suspected that he was from the South.) The letter was also funny and—unique among donor correspondence—modest. The author was obviously a smart man, but he didn't show off. He referred to himself only in order to deprecate himself. But what struck me most about the e-mail was how romantic it was. Not romantic in the moon-June-for-you-I-swoon sense but romantic in the sense of romantic poetry—filled with a childlike sense of wonderment, possibility, and love.

I was smitten by Donor White. Still, I was on guard for hoaxsters, and on second reading I realized that "Donor White's" note lacked identifying details. He was specific about his family history, but since I didn't know his family, he could have been making it all up. And when he described the Repository, he included no fact that he couldn't have gleaned from reading my articles about it. So I replied to him with a curt e-mail quiz. I demanded information that only Donor White could know: Beth and Joy's real first names (which he had been told) and minutiae about his ancestors that he'd revealed in his correspondence with Beth seven years before.

Donor White took the quiz in the right spirit—he was a cautious man, too. He aced it. So on June 13, I called Beth and told her that Donor White had found me. She could barely speak. (In our excitement, we started whispering again.) I gave Beth his e-mail address. I mentioned that as far as I could determine, this would be the first time an anonymous sperm donor and his child had ever met.

I felt oddly ambivalent about introducing Donor White to Beth and Joy. My life as the Semen Detective had been straining my conscience. I had decided that as long as I was careful not to identify those who wished to remain invisible, I would be doing no wrong. Failure had presented one kind of dilemma. Donor Fuchsia's insistence on privacy, for example, meant that the eight kids of his I had found would never know about their father, even though I did. There was also a pair of half siblings who, because one parent asked for privacy, wouldn't ever meet each other or even know about each other. I hated having to keep these secrets, but I had no choice.

But now that Beth and Donor White were on the cusp of meeting, my conscience was muddier. Success presented a different, more demanding moral challenge. I worked my way through the dilemma. First: Beth and Donor White. Was it okay for them to meet? Beth was a sensible, well-meaning adult. Donor White seemed a sensible, well-meaning adult. What was the harm in letting them talk to each other? None.

Okay, second: Joy and Donor White. Was it okay for them to meet? Joy was only twelve years old. Meeting Donor White might upend her life, yet she wasn't permitted to stop it from happening.

Her mother would decide for her. I rationalized this: Parents always make decisions—even traumatic ones—for their children. That is a parent's job. I had to trust Beth to do right for Joy, just as I would expect others to trust me to do right for my kids. So that was fine.

Now the third and hardest problem: Joy and her father. Was I doing her father wrong? I had never talked to Joy's dad—Beth's ex-husband—and didn't even know his name. It's true that Beth, not I, would decide whether Joy got to meet Donor White. But that was a technical distinction. I knew that what I was doing could alter Beth's ex-husband's life, perhaps for the worse. I was helping Joy connect with a second father who might compete with him. Was this justified? Perhaps not except in the utilitarian sense: the potential benefit to Beth, Donor White, and Joy outweighed the possible pain it might cause the father.

Donor White and I struck up a lively correspondence, and it was immediately clear that he was just as sweet a man as his initial e-mail suggested. It didn't make much sense that Donor White and I got along as well as we did. He was a Southerner by birth, a Californian by lifestyle, a scientist by vocation, and a Republican by sensibility. I was none of the above. But Donor White reminded me a bit of my father, and not just because they were both scientists in their sixties. They possessed the same balance of rationality and kindness. Donor White gave any question I asked him two answers, a logical one and a soulful one—sometimes they matched, sometimes they didn't. Like my father, Donor White could hold in his head the incompatible demands of rationality and irrationality, of facts and love.

Despite our warming friendship, he remained something of a riddle to me. I was sure he was Donor White and that he had contributed sperm to the Nobel bank, but beyond that, I hit a brick wall. He was cagey about his identity. Unlike every other donor I had talked to, he didn't tell me his real name, where he lived, or what he did for a living (beyond "scientist"). He used an untraceable e-mail address—an e-mail address that had never shown up on the Internet with a name attached to it. He wouldn't give me his phone number, and he wouldn't call me, so I never heard his voice. We communicated only by e-mail. He had shared his true identity with Beth, she told me, but he had asked her to keep it from me. It wasn't that he didn't trust me, he insisted, he was just worried about any number of circumstances where the pressure on me to surrender his name could be intense. If I didn't know it, I would have nothing to say.

Eventually, Donor White did feed me a few crumbs of information. He admitted that he lived near San Diego, and he gave his first name—"Roger," let's say.

I begged him to let me visit him in San Diego. He reluctantly agreed and laid out his conditions. He would bring some documents to show me but would not carry anything that identified him, not even a driver's license. He set a meeting place, a San Diego hotel at ten on a Saturday morning. He would be sitting in the lobby carrying a brown satchel. He wanted privacy, so I was to reserve a room where we could talk.

The day came. When I arrived, there was only one person sitting in the lobby, an old man. Roger's e-mails were so full of youthful enthusiasm that I had forgotten, as we had corresponded, that he was nearing seventy. He had been recruited as a donor after his fiftieth birthday, so he was twenty years older than most of the other Repository donors I'd met. Roger stood to greet me. He was six feet tall, with a pot belly, and he had the presence of an even larger man. That was because his face was so big and so round: a shiny full moon of a face. He was not quite handsome—his features were too fleshy for that—but he looked . . . nice. His cheeks were large and sagging down into jowls. His coloring was rich and red. The "dark brown" hair listed in the donor catalog had turned gray but remained bushy. The anchor of his face was a strong, appealing nose: it almost seemed to be three noses, the bridge and each nostril were so massive. (Seeing this distinctive nose was what made me certain he was Joy's father: she had that nose, too, in a more delicate, girlish form.

The nose knows.) Roger's eyes were blue and cheery behind a clunky pair of bifocals. He wore high-water pants and a striped shirt that looked as if it should have a pocket protector but didn't.

Roger welcomed me with a strong handshake, a smile, and a southern gentleman's grace. He said it was "pleasing" to meet me. "Pleasing" was the linchpin of his vocabulary. The word occurred over and over again in his conversation. When something mattered to him, he said it was "pleasing." Getting letters from Beth years ago, that had been "pleasing" to him. Seeing pictures of Joy, that was "pleasing." And the chance to meet Joy, well, that would be "very pleasing."

When we got up to the hotel room I had booked, Roger unsnapped his briefcase and stacked a sheaf of manila folders on the table. Every folder was labeled, and they were arranged in a precise order. Methodically, he worked his way down the pile. Some contained original documents or private documents that I was allowed to examine in the room but could not take home. Other folders contained only photocopies that were mine to keep. Where the photocopies were blurry or cut off, Roger had neatly printed the missing words. His archive was both idiosyncratic and elaborate: he had brought everything from a copy of Graham's original 1963 agreement with Hermann Muller to the Christmas cards that the Repository sent donors to the program from Graham's funeral.

As Roger talked about the documents and the Repository, what struck me most about him was not that he was a precise scientific man, though he clearly was. What struck me was that he was a precise scientific man who had been bonked on the head by a miracle. Roger seemed physically unsettled by the discovery of Joy. One of the first things he said was "I am a technical guy. I believe in *facts*. But so many strange things have been happening to me! Finding Joy again, there is just no scientific basis for that. I have never believed in destiny, but now I think there must be something to it."

He told me how he had become a donor to the Repository. He

began — ever meticulous — at the beginning. He pulled out a photograph of himself as a toddler. He looked as Joy did in her baby photos, but perhaps just in the way all cute little blond kids look alike. Roger had grown up in small-town Alabama in the 1940s and '50s. His father, Roger said, had been a machinist, and though he had only finished high school, he was a numbers genius. His dad could multiply huge numbers in his head, and keep a running count of the number of letters in Sunday's church sermon. His dad had died when Roger was a teenager, leaving Roger a kind of surrogate father to his younger sister, herself a math prodigy. Roger was no genius child, but he was dogged. He earned degree after degree in chemistry from state universities. When he graduated, the chemical industry was booming, and he easily found work, first in Texas, then in California.

The clues he dropped in our conversation — Alabama, chemicals — were enough, combined with a few other hints, for me to identify him later using an online database. As he had told me, he was a successful, but not famous, scientist.

Roger worked with the intensity of a poor kid made good. He scarcely had time to date, and he didn't marry until his thirties, when he met Rebecca, a colleague several years his senior. Roger had always been a family man: he was devoted to his mom and sister, and he wanted to be a father. Rebecca already had a child by a first marriage, so when Roger and Rebecca couldn't conceive, he assumed that he was sterile. They talked about adoption but never got around to it. He was working too hard.

His professional success seemed satisfaction enough. He earned four patents. Scientific papers and technical reports piled up by the dozen. He leapt from one snazzy company to another. He was always in demand, not because he had the best brain but because he had a pit bull brain. Roger grabbed problems in his teeth, shook them like mad, and wouldn't let go till he had broken them. He worked twelve-hour days, week after week, year after year. It took a toll on him. But he pretended it didn't.

In July 1984, two women visited Roger, unannounced, at his laboratory. They told him that their names were Julianna McKillop and Dora Vaux and they worked for the Repository for Germinal Choice. Had he heard of it? He had heard of it. He was dumbfounded. He had enough presence of mind to shut his door and ask his secretary to hold his calls. Roger still didn't know how they had selected him— he suspected that a former colleague had tipped Graham off to him. Roger was a great candidate for Graham's post-Nobel bank—a Renaissance scientist. He was plenty smart, but he was also big, friendly, hardworking, and very athletic. Julianna and Dora told him he could do the world a great favor by donating to the bank and preserving his wonderful genes. "I listened, without saying much, mainly because of being virtually speechless," Roger remembered. "I would never have thought about such a thing in my entire lifetime. Not wishing to be rude, I told them that I would need to think about this myself for some time and then speak to my wife before getting back to them." He reminded them that he was over fifty and had never fathered a child. They told him not to worry about that yet.

Roger was inclined to reject the invitation, but he was a deliberate, contemplative man. So he turned the idea over in his head for months. Every few weeks, he told me, Julianna would mail him a testimonial from a happy mother or a glossy pamphlet or a videotape of a news program about the Repository. He was not swayed. Graham's eugenic ambitions did not move Roger; he thought DNA was far too fickle to guarantee the superkids that Graham sought. Roger decided to turn Graham down. He wrote the encounter off, and filed it in his head under "strange experiences."

Then, for practically the first time in his life, Roger allowed himself to be interrupted—and that's exactly how he thought of it, as an interruption—by fate. He rarely dreamed, and when he did dream, it was always in black and white. But about six months after Julianna and Dora's visit, he had a Technicolor dream. In the days before the dream, he had been researching his family history. Roger had long known that his great-grandfather had died in the Civil War, fighting for the Confederacy. He had just discovered that the great-granddad had fathered his only child right before he died in the war.

So this was the dream Roger had:

I was sitting on the edge of an open field with my back against the trunk of a giant oak tree. It was a beautiful day and monarch butterflies were flitting about all around me, when some distance away the outline of a man could be seen coming out of the field toward me. There was a bright light at his back that blinded me until he came close enough to fall within the shade of the tree, at which time I immediately knew who he was before a single word was said. While no photograph of him existed, I knew that this poorly dressed man was my great-grandfather from the Civil War, because he looked exactly like a composite of my father and grandfather.

Without any introduction, he spoke to me as follows: "Most of my friends volunteered at the first opportunity to enter the war. I was newly married and waited until there was danger of being conscripted before joining up. Because of that I had a son . . . which is the only reason that you and all of those known to you having my name ever had a chance at life. You now have that same opportunity."

The dream changed his mind. He called Julianna back and agreed to take a physical and give a semen sample. To his surprise, his spermatozoa were both numerous and lively. He also passed the Repository's medical exam. Julianna code-named him Donor White

#6 and wrote a catalog entry describing him as "a scientist involved in sophisticated research" with "good features, good presence."

For Roger, becoming a sperm donor was an act of moral purpose. He had committed to help couples who needed him, and by God he would not disappoint them. Once he determined to do it, he did it with the care that he gave to everything that mattered. He wasn't paid a penny, but Roger made himself as passionate about donating sperm as he was about running chemical reactions. He learned to process sperm at home, how to preserve it with extender solution, pipette it into tiny vials, top each with a white screw top (hence, Donor White), and freeze them in liquid nitrogen. Every few months, Dora Vaux would leave an empty liquid-nitrogen Dewar flask on his front porch and collect the Dewar he had filled up. It felt productive to Roger, and it felt right.

Roger also insisted that donating sperm had to be an act of love. In the peculiar transaction that is sperm donation, donors and sperm bankers leave a lot unsaid. They don't talk about the fact that, at its heart, sperm donation is a furtive, in-a-closet-with-a-porno-mag process. It's lonely, and—trust me, because I have been through it—skanky. It is exactly what it seems to be: jacking off. That's why sperm banks avoid telling donors exactly what they're supposed to do. Instead, they couch it in euphemism: "donation," "collection," and "processing."

But Roger rejected the sleaze and the furtiveness. Sperm donation, he determined, "need not be a solitary activity." Roger had married well: Rebecca tolerated, even encouraged, Roger's sperm donations. She had friends who had suffered through infertility and thought it was a good deed to assist other couples who longed for babies. She was willing to help. So when Roger and Rebecca made love, he collected the sperm in a special condom and saved it for the bank.

At first, Roger said, he seemed to be shooting blanks. Donor

White sperm wasn't getting anyone pregnant. Finally, in 1986, Dora called him: the first Donor White baby had been born. Soon the White babies were arriving at a rapid clip—one every couple of months. By 1990, he had fathered a dozen kids. By 1991, nineteen of them.

Why did Roger know the number? Because he kept records. He examined data for a living; this was data. Roger opened one of his folders and handed me a handwritten graph. The Y-axis read, "Conceptions for Donor White." The X-axis had the year. He unfolded another graph, which charted how many babies had been born at the Repository—from all donors—when each manager was in charge. I asked him how he'd collected the numbers. He said that he and Rebecca had struck up a friendship with Dora. When she delivered a tank, she would drop in for tea and spill secrets. Whenever a White baby was born, she told him the birthday and sometimes the first name.

With his typical orderliness, Roger also took the occasion of being a sperm donor to make himself a student of fertility. He read scientific papers about it, and once, when he encountered a curious fact in the literature—that women married to older men have disproportionate numbers of boys—he saw an opportunity to contribute to fertility research. Did the same anomaly exist with donated sperm? Since Repository donors tended to be older than the mothers, you could check if their older sperm also tended to produce boys. He wrote to Graham requesting the bank's data on sex of offspring. Graham never responded.

But Roger's fascination was only incidentally scientific. He was enthralled by the numbers because the numbers represented life. Each number was a child—*his* child. The older he got, the more he thought about his distant kids. He did not ask the obvious, vain question about them: Are they like me? (That is what all the other donors asked.) No, he asked the questions a father asks: Are they happy? Are they healthy? What are they going to do with their lives? He thought of all the paternal care and advice they deserved and how he couldn't provide it. That depressed him. All he could do was think about them.

Since he didn't know what was actually happening to his children, he had to imagine it. Roger removed a little black notebook from one of his manila folders. He leaned over conspiratorially—though we were alone in the room—to show it to me. Each page of the notebook listed a birthday. Often there was a first name written below the date. Every time Dora informed him of a birth, Roger recorded it here. A surprising number of parents mailed baby photos to the Repository as thanks, and Dora sent them on to Roger. He had a dozen of them. Each of them was backed by a sheet of cardboard, sealed in its own protective plastic bag and inserted into the notebook by the corresponding name.

Roger leafed through the black book. He pointed to a boy named Avi Jacob. "He's probably Jewish," Roger said, smiling buoyantly at me. "So, David, you are not the only one here who has a Jewish child!"

On the next page were twins, a boy and a girl. Roger had two pictures of them: one at four months, a second at eleven months. Roger noticed what I didn't, that the boy was smaller than his twin sister in the first picture and bigger in the second. "Look how much he is growing!"

Next came a darling curly-haired boy: "He was just four pounds when he was born. They were worried. But now look at him!"

Then Roger showed me a large, curious picture of a baby boy in a bathtub. The bathtub was at the top of a tower in the middle of a forest. "Dora said his parents worked as fire watchers," said Roger. What's his name? I asked. It was not written on the notebook page. Roger shrugged, then grinned. "Dora didn't tell me. But I even have names for those whose names I don't know. So I call him 'Watchtower Boy.' Or sometimes 'Boy in Tub.'"

Roger closed the book. "Maybe Dora should not have shown me

these pictures, but I am glad she did. These children are very pleasing to me," said Roger. And here, "pleasing" meant something very much more than pleasing.

He handed me several letters. They were thank-you notes sent by "White" mothers that Dora had passed on to him. One read, "Every mother believes hers to be the most special baby ever born, but mine truly is." Another read, "You have given us the greatest of all gifts, more precious than anything money can buy, and changed the way I feel about human nature. You are an unseen but not unfelt member of my family."

Roger commented that Joy was not the first of his children whose identity he knew. In the late 1980s, Dora had, as usual, told him the first name of the latest White child. It was an unusual name, "Jeroboam," let's say. A couple of years later, Dora mentioned to Roger that the boy was about to have a baby sister, also by Donor White. The next week, Roger happened to see a birth announcement in his local newspaper. The announcement mentioned that the new baby girl had an older brother, Jeroboam. Roger was thrilled. He looked up the family in the phone directory and realized they lived only a mile away. He took to running by their house on his morning jog, especially on Christmas morning and on the kids' birthdays, which Dora had told him. Sometimes he would see Jeroboam playing in the yard. Once, when Roger jogged by during the boy's birthday party, Jeroboam shouted to him, "Hello, Man in the Hat." Roger stopped and said, "Happy Birthday." Soon after, he made a plan to meet the boy. He ran by the house carrying a beach ball and rang the doorbell. He pretended he had found the ball on the street and wanted "to give it to the little guy who sometimes waved at me when I jogged past." The mother, suspicious, stayed inside the house. Roger had to shout his story at her through the locked front door. She thanked him and asked him to leave the ball on the porch. Later, he saw Jeroboam playing with the ball. It was ten years since then and he had never spoken to the boy—or his sister—again. He knew that he should not.

As Roger showed me his notebook and told me his stories, I felt heartsick for him. This was what happened when a deliberate man with a pure soul became a sperm donor. He had tracked his children because he felt he must. It was the closest he could come to being the father they deserved. He knew he would never—and could never—interfere in their lives, and that agonized him. The Repository children—at once his and not his—were excruciating for Roger. In October 1991, he wrote an article for *San Diego Woman* magazine, under the pseudonym R. White. (It was an early draft of the article that Dora had shown Beth when she was shopping for donors.) The magazine article described the sperm bank kids as being a *kind* of comfort to him, because at least he had the satisfaction of passing on the genes of his ancestors (*not* his own genes, he was too modest to say that). He also joked about his fertility: "It is really beyond the imagination of a fifty-seven-year-old man, who thought only five years ago that he might be infertile, to realize that he now has enough boys for a baseball team (with one extra to umpire) and enough girls for a basketball team."

But mostly Roger's article was elegiac because he knew that he would never see his children. "Fathering children anonymously is somewhat akin to producing paintings that to you are beautiful and priceless, but doing this with the understanding that when they are finished they must be given away and likely never seen again." This was a haunting image. Whenever someone asks me what it's like to donate sperm, I quote them that passage. For Roger, every time he learned of another birth, he felt pride and he felt loss: every donation was an act of loving-kindness and of pain.

Roger stopped donating shortly after he wrote the article. He had fathered nearly twenty children. The Repository professed a limit of twenty per donor, but Graham had urged him to continue anyway. Roger thought he shouldn't. His sense of obligation made him stop.

Becoming a semifather had another effect on Roger, one he hadn't expected. The children—the children he couldn't ever know—made him feel he had wasted too much of his life on work. In the mid-1990s, he retired, full of regret for all the hours squandered at the office. Whenever I mentioned to Roger that I was taking a business trip, he rebuked me for spending time away from my wife and kids. As we were chatting in the hotel about something else, he suddenly grabbed my arm: "I was a workaholic. I regret it. I missed time with my family. Don't *ever* do that, David."

All of the nineteen White children occupied Roger's thoughts, but the one who occupied them most was Joy. Their 1991 meeting—when Beth had dropped baby Joy at the Repository so Roger could see her—had affected him profoundly. As we talked, Roger recalled the visit as if it had been the day before: When Dora had called, he and Rebecca had raced over to the office. Joy, he said, had immediately held out her arms to him, inviting him to pick her up. He had been astounded at how much Joy looked like his sister had when she was an infant. For half an hour, Roger had held little Joy on his lap while Rebecca and Dora snapped Polaroid after Polaroid. Roger had kept these, of course, and he slid them out of a folder and showed them to me: Joy was wearing a light blue romper, with a tiny lace collar. Roger, then in his late fifties, stared at the camera with a stunned expression of wonder, fear, shock, and gratitude. I recognized the expression from family snapshots of me after my daughter's birth. It was the look of all first-time fathers.

Eventually, Roger said, Joy had squirmed off his lap, scooted over to her stroller, and tried to pull herself into it. She couldn't manage it on her own, so Roger had lifted her up and placed her in the seat. Joy had been annoyed at the assistance. She had slid herself off the stroller and tried again till she managed to climb in by herself. This also reminded Roger of his beloved sister, who could never tolerate being helped. Roger had turned to Rebecca and said, "We really have ourselves something special here." After forty minutes, Dora had told Roger they would have to leave. Roger had kissed Joy and said, "I will see you again." She couldn't understand him, of course, and he had known it was impossible, that he could never see her again.

Roger's reaction to the visit showed why sperm banks forbid meetings between donor and child. Roger was supposed to be a detached, anonymous donor, a source of fine DNA and nothing more. But this visit had bonded him to Joy. She was no longer simply "donor offspring." She was his *daughter*. When he got home from seeing Joy at the Repository, Roger wrote a letter to her, a letter he could not send, since he did not know her last name or address. In it he told Joy that the visit had been enough to make him feel like a father. He had kept the letter since then, locked away safely at home.

After meeting baby Joy, Roger tried to accept that he would never hear from her again. But a few years later, he started receiving occasional photos of Joy, sent by Beth and passed on by the Repository. When the Repository permitted the correspondence between him and Beth to accelerate in 1995 and 1996, he was ecstatic. He wrote much less than he wanted, for fear the Repository would stop the letters if he was too avid. Roger framed the pictures of Joy that Beth sent. He hung a collage of eight snapshots in his living room and put the photo of Joy skiing—blurry, unrecognizable—above the mantelpiece. He was proudest of a picture of Joy sitting on the lap of a Confederate soldier. Dora had told Beth about Roger's dream of his Confederate great-grandfather. When Joy was two Beth had taken her to a Civil War reenactment near their town and had a photo taken of Joy with a Confederate soldier. Beth had thought Donor White would appreciate the coincidence. Roger showed me a photocopy of the picture of Joy and the reenactor. The soldier in the picture, he said, looked just like his great-grandfather in the dream.

Donor White was heartbroken when the correspondence with Beth ceased in 1997. He didn't know why it had ended, because the Repository didn't send him the explanation that it sent Beth. Then, two years later, he received a letter from the Repository's medical director, announcing that the sperm bank was closing. Roger feared that once the bank closed, his faint remaining hope of finding Beth and Joy again would vanish. He started improvising. He reexamined the photographs Beth had sent him. Printed on the back of one—and overlooked by the Repository—were the name and address of a Pennsylvania photo studio. Then he studied Beth's correspondence and noticed that she had used personal stationery for the first letter she had sent, way back in 1991. That stationery was embossed with a lily. It occurred to him that her last name might be "Lily." He searched for an Elizabeth Lily in every state. He found only one, and she lived in Pennsylvania, near the town where the photo studio was located. He immediately sent a letter to her. It was extremely cryptic. In it, he wrote that a person named Beth had befriended him through a third party and that third party was no longer available to pass on information. If Beth Lily was the Beth who had befriended him, then she could write to him at such and such an e-mail address. He added that he had only the very best intentions toward Beth and that if she was the wrong person, she should please discard this strange letter. He put no return address on it. No one but Beth herself would have had the foggiest idea what he meant.

Roger had guessed wrong, and he heard nothing back from Elizabeth Lily. He resigned himself to the loss and to a life without children. He filled his retirement with studying Civil War history and investigating his family tree. In 1999, he finally told his eighty-nine-year-old mother about his sperm bank kids. She took the news very cheerfully. When she died a few months later, Roger was relieved that he had shared the secret with her.

In early 2001, right at the time I published Beth's plea for Donor White in *Slate*, Roger fell seriously ill. I had feared his yearlong silence might have meant he was dead, and, in fact, he almost was. Although he slowly recovered, he wasn't the same man. The illness had left him with permanent complications. He had shed thirty pounds, wasting down to skin and bones. He couldn't jog anymore. He wasn't even seventy yet, but he could barely manage to shop for groceries now. A great specimen of a man, he had become weak, his athletic body slack. If his body was broken, his spirit was worse. "I thought, 'Life is not worth living. If I die now, that's fine.' " As he slowly recuperated, Roger berated himself for having spent his life in the lab. He belittled his scientific accomplishments; his patents meant nothing next to his nineteen children—the children he would never know.

In June 2002, a friend e-mailed him a Web page with a link to an Internet search engine, Alltheweb.com. Roger was a novice Web surfer. He had never even heard of search engines. He went to Alltheweb, and the first thing he typed in was "genius sperm bank." The first entries were for my *Slate* series on the Repository. He scrolled down the page. Halfway down he saw this listing: "A Mother Searches for 'Donor White.' " He was stunned. He e-mailed me that day.

June 16, 2002—four days after Roger wrote to me—was Father's Day. Roger logged on to AOL. "You've got mail." A message from an unfamiliar address was waiting. He opened it. It was from Beth. It began, "Happy Father's Day!"

CHAPTER 9
MY SHORT, SCARY CAREER
AS A SPERM DONOR

Retrieving vials of frozen sperm from a liquid-nitrogen storage tank. *Courtesy of California Cryobank*

I realized that I needed to donate sperm, too. Not because I wanted to, quite the contrary. My son had been born in early 2003, so I was the father of two children, which seemed more than enough on most days. My lack of desire to donate is why I felt obliged to do it. No matter how often donors explained themselves to me, sperm donation befuddled me. I nodded and smiled at them as they listed their reasons, but my own brain was snickering. Why had they subjected themselves to such inconvenience? To such embarrassment? Roger had made the most compelling possible case for sperm donation to me, and I still didn't get it. I had to find out for myself what I was missing.

I dutifully informed my wife about the plan to donate. "No way," Hanna said. I argued that it was all in the name of research. She was unimpressed. I promised that I would stop the sperm bank before it could sell my sperm. She said she didn't believe the bank would make such a deal. I swore that there was no chance they would use my sperm. I begged, which was not a pretty sight. She relented.

These days, all sperm banks recruit customers and donors through the Internet, so I cruised the Web and quickly found an application for the big local bank: Fairfax Cryobank, located in Washington, D.C.'s, Virginia suburbs. Fairfax Cryobank is to sperm banking what Citigroup is to real banking. It has branches in four states and Canada. The sperm bank itself is only one small division of a full-service fertility business, the Genetics & IVF Institute.

I completed Fairfax's online application in a couple of minutes—it asked for the barest minimum of information, hardly more than my name and address. A week later, the mailman delivered a brown envelope with no name on the return address. Sperm banks, like pornographers, keep everything on the down low: mail invariably arrives in discreet envelopes. Bank staffers dislike leaving phone messages, but if they must, the message is almost incomprehensibly vague: "This is Mary, from Fairfax"—Fairfax what? I would ask myself—"We'd like to talk to you about your recent inquiry. Please call us at . . .")

The brown package from Fairfax contained the full application, an eighteen-page slog. I trudged through the physical data: age, hair color, height, weight, blood type. Then I dragged my way through the biographical section: educational history, profession, musical talent ("None," I wrote proudly), athletic abilities, hobbies. Then I labored through the medical questionnaire: alcohol use, tobacco use, drug use, tattooing history, how well I sleep, how well I eat, what medicines I take and why, what bones I have broken, whether I exchange sex for money, whether I had used intranasal cocaine in the preceding twelve months. I listed three generations of familial mental illness and felt my own ticker skip a beat when I wrote that all my male ancestors on both sides of the family had died young of heart disease. I declared that I wasn't a carrier of Gaucher's disease, Fanconi's anemia, Niemann-Pick disease, Canavan disease, or thalassemia, although I had not the faintest idea what those illnesses were. I had to check off whether I suffered from any of an endless roster of symptoms—hoarseness, warts, blood in stool, goiter, tingling, dizziness, fainting, convulsions, seizures, fits, shaking, tremor, numbness. By the time I was done, I was suffering from several of them. I was asked sixteen ways to Sunday if I inject drugs or have sex with other men. I agreed to submit to an HIV test. Finally, I reached page eighteen, which was the scariest of all: "I agree that I release all rights, privileges, and disposition of my semen specimens to Fairfax Cryobank." Hanna is going to kill me, I thought, and then I signed it.

According to the application, if my written application made the cut, I would be invited for an interview, where I would "produce" a semen sample for analysis. If that were satisfactory, I would return for more semen analyses and a physical. Only if I passed those would I qualify as a donor.

I mailed my application to Fairfax and waited. And waited. And waited. After two months, I was furious. How dare they ignore my semen? That semen had produced two healthy children! That semen had graduated from an Ivy League college! That semen had run a marathon! Then my rage turned to worry: Did Fairfax know something I didn't about my health? Was my future that bleak? Was all that heart disease really so bad? Suddenly I found myself desperate to be chosen.

I had finally given up on Fairfax and applied to a bank in New

York when I received an e-mail from Amanda, who identified herself as Fairfax's laboratories coordinator. She invited me for an interview. She noted, oh so casually, that I would have to furnish a sample on the premises.

I made an appointment for the following Monday. Fairfax Cryobank was located beyond the Washington Beltway in The Land of Wretched Office Parks. The cryobank was housed in the dreariest of all office developments. To call the building anonymous would insult anonymity. The ugliness may be intentional: a sperm bank doesn't want to draw attention to itself or its visitors. I hunted through the first-floor corridors, past the mysterious "microsort" room and "egg donor" facility, searching for the sperm bank office. I saw an open door and peeked in. It was the room where Fairfax stored its extra liquid nitrogen. There were four tanks inside. They looked like fat silver men. The frozen sperm itself was under lock and key elsewhere. I wanted to see that room, to see the tanks that held Fairfax's thousands of vials, each vial filled with millions of spermatozoa. Finding that room would be like the scene in the science fiction movie when the hero accidentally discovers the warehouse where the "friendly" aliens are freezing the millions of humans they have secretly kidnapped for their terrible experiments.

Finally I located the door marked "Cryobank" and walked into an uncomfortably cramped waiting room. A couple—not a young couple—was sitting there. They looked up, startled, when I entered. We half smiled uncomfortably at each other. All of us instantly recognized the awkward situation. They were there to buy sperm; I was there to sell it. We had each accidentally looked through a window into a world we did not want to see. I was sure the couple was thinking, *That guy* is a donor? The hell with this place, let's go to Sperm World instead.

I flagged down the receptionist, who assumed I was a customer, too. When I explained I was there to see Amanda about donating, she was chagrined. I wasn't supposed to be there. I had apparently come in the wrong door. Amanda was summoned from her office and hustled me into the back, out of sight of the couple.

Amanda led me to her office, a cozy place with wedding pictures and prints of sailing ships. She checked my driver's license to make sure I was who I claimed to be. Then she pulled out my application and began reviewing it with me, line by line. In *tone*, it felt like a job interview with a vice president for human resources. In *subject*, it was rather different. "Okay, so you live in Washington, great. And your blood is B positive. You sure of that? No? That's okay, we'll check it. Hmm, so your family is from eastern Europe. Do you know exactly where? Can you check?" She noticed I was married and asked if my wife knew that I was there. I answered, "Of course. Don't all wives know?"

Amanda acted as though this was very funny and said, "A lot of donors are married and don't tell their wives." (And *these* are the guys you want to father children?)

She asked me where I had gone to college. I said "Harvard." She was delighted. She continued, "And have you done some graduate work?" I said no. She looked disappointed. "But surely you are *planning* to do some graduate work?" Again I said no. She was deflated and told me why. Fairfax has something it calls—I'm not kidding—its "doctorate program." For a premium, mothers can buy sperm from donors who have doctoral degrees or are pursuing them. What counts as a doctor? I asked. Medicine, dentistry, pharmacy, optometry, law (lawyers are doctors? yes—the "juris doctorate"), and chiropractic. Don't say you weren't warned: your premium "doctoral" sperm may have come from a student chiropractor.

After a few minutes discussing the application, my attention wandered. I gazed absently at Amanda's screen saver, a soothing blue-and-white pattern just over her shoulder. After a few seconds, I noticed that the white pattern wasn't a pattern. It was a school of tiny sperm, tails waving jauntily as they motored across the screen. I took a second look at the mouse paperweight on Amanda's desk. It wasn't a mouse. It was a cute little sperm.

Such goofiness was, I came to discover, a hallmark of modern sperm banks. Fairfax hands out pens on college campuses that ask, "Why not get paid for it?" When I visited California Cryobank, the director of public relations gave me a T-shirt depicting swimming sperm. Around the sperm ran a circle of text that read "Future People" in a dozen different languages. California Cryobank distributes pens, too. They're floaty pens, with a little plastic sperm swimming up and down, up and down.

Anyway, back to Amanda. At this point I am obliged to point out that Amanda was cute. In fact, she was distractingly cute. She was thirty, I'd guess, and looked Latina. She smiled all the time, a sexy, gleaming smile, and laughed when I made even the lamest stab at a joke. She leaned across her desk toward me as we talked.

Rule number one of sperm banking: The people who recruit donors are invariably women, and they are invariably good-looking. I suspect—no, I am sure—that this is deliberate, to get donors excited to join the Fairfax team.

Yet Amanda's sexiness presented a kind of paradox. The chief activity of the sperm bank—its entire purpose—is masturbation. But my interview with Amanda was actually designed to *desexualize* what I would be doing. The goal of the interview seemed to be to eliminate the embarrassment that men feel about masturbation by replacing it with tedium. After the endless review of my application, Amanda walked me step by countless step through the qualification process—if my sperm count was above such-and-such a number, I would make the next round. There would be blood tests for gonorrhea, syphilis, hepatitis, and lots of scary diseases I had never heard of. They would give me a renal ultrasound. My sperm would again be counted, frozen, thawed, and recounted. Its motility—how well it swims—would be tested and retested. Only then would I finally be

admitted as a donor—and even that was contingent on passing regular blood tests. Amanda then listed what I would be required to
supply to the bank if I qualified: baby photos, an audio CD about
myself, essays on such topics as "What is your most memorable
childhood experience?" and "What is the funniest thing that ever
happened to you?"

After that, Amanda held forth enthusiastically and at great
length about money. "You will get paid $50 per usable specimen, for
starters. Then you will get $5 for every vial from the specimen. The
average is ten to fourteen vials per specimen. When a vial is released
from quarantine after six months, you will get another $5. So the av-
erage payment is $209 per deposit." She paused. "Now, this is ordi-
nary income, but we don't do withholding. We send checks twice a
month, but later we will just give you a check every six months. We
will send you a 1099 form at the end of the year."

Amanda had managed to take a mysterious and sexual and profound process and make it sound exactly like . . . a job. I considered
asking her about the 401(k) and dental benefits.

Finally, it was time for the money shot. She led me next door to
the lab, where three women in lab coats were chatting about their
weekends while gazing at sperm samples under microscopes. They
ignored me. When I became a regular donor, Amanda said, I would
come straight to the lab to collect a sterile cup and a labeling sticker.
She handed me a cup. Amanda pointed to a small incubator—a
warm metal box—where I would put the "specimen" when I was
done. Next to the incubator was a pile of plastic sachets; they looked
like the mustard packets you get with a deli sandwich. "That's KY
jelly," she said. "It's nontoxic for sperm. Still, just try not to get it, you
know, on the *sample*."

Amanda escorted me back down the hall to a donor room. Fairfax has two of these—sometimes known in the trade as "blue rooms"
or "masturbatoriums." The room was really no more than a large

closet. A dingy beige love seat was pushed against the far wall. An erotic print hung on the wall above the sofa. It was a painting of a woman from behind; she was wearing some diaphanous lingerie. It was pretty sexy, to be honest. On another wall were a clock, a sink, and a cabinet. Amanda handed me a pen and told me to write the time of ejaculation on the cup when I was done. She turned on the taps and instructed, "Wash your hands with this antibacterial soap, and dry them *well*. Water is toxic for semen."

"Here's the exhaust fan." She flipped a switch by the door, and a buzzing noise covered the room. She opened the cabinet again. "And here are the magazines." She handed me a stack of *High Societys*, *Gallerys*, and *Playboys*, all, shall we say, well thumbed. "Fairfax Cryobank" was scrawled on the cover of each of the porno mags. Amanda, who did this routine several times a day, seemed unfazed. It was just a commercial transaction for her. I pretended I was unfazed, too.

She gave me the phone number for the chief lab technician and told me to call the next day to find out whether I had a high enough sperm count and whether my guys had survived freezing and thawing. "Now, of one hundred men who apply," she said reassuringly, "we only interview twenty or thirty. And the vast majority of those— even men who have their own children already—end up being disqualified by sperm count. So *don't feel bad* if you don't make it." She thanked me for coming in. She flashed me one more gleaming, sexy smile, closed the door, and locked it from the outside.

The next few minutes passed as you would expect and are none of your business.

When I was done, I walked my cup down the hall to the incubator. I tried to catch the eye of one of the technicians, to ask if I could take a sperm paperweight as a souvenir. None of them looked at me.

The next morning, I called the chief lab technician. "I was

about to call you," she said. "I have some good news. You passed the freezing and thawing. We want to make arrangements for your second trial specimen—that is, if you are still interested."

I flushed. I couldn't resist asking "So what were my numbers? What was the count?"

"Your count was about 105 million per milliliter. The usual is around fifty to sixty million. So you are well above average."

I grinned—105 million! I considered breaking my promise to Hanna, and continuing as a donor. I was, after all, "well above average." I started to make an appointment for my second deposit, then thought better of it. Hanna was right: Who knew what they were doing with my sperm? The longer I kept up the charade, the greater the possibility that my sperm would end up in the wrong hands (or wrong uterus). So I told the tech I needed to check my schedule and would call back. I didn't call back. She phoned me again a few days later and left a vague message. I didn't return it.

I was not much closer to wanting to be a donor than I had been before I started, but I was closer to understanding why someone else might want to do it. In the abstract, donating sperm had seemed fundamentally silly. But actually doing it was seductive. I had been accepted by the ultraexclusive Fairfax Cryobank! My sperm was "well above average"! My count was 105 million! What's yours, George Clooney? Amanda, lovely Amanda, had asked for my help. The women of America—barren, desolate, desperate—needed *me*. They yearned for my B-positive, brown-eyed, six-foot-one-inch, HIV-negative, drug-free, heart-attack-prone, only slightly mentally ill sperm. And what kind of selfish monster was I to deny it to them?

From the vantage point of today's fertility-crazed America, where people talk about their fertility specialists the way they used to talk about their plumbers, where every woman is either making a baby through in vitro fertilization, donating an egg so someone else can,

carrying a surrogate baby for her daughter or mother or rich neigh-
bor, or seeking to adopt her lesbian partner's hormone-spawned
sperm-bank triplets, where getting pregnant the old-fashioned way
seems not merely old-fashioned but slightly foolish, it can be hard to
remember that infertility used to be a badge of shame, that the only
fertility choice women used to have was "Blue eyes for the donor or
brown?" and that they were supposed to be grateful even for that.
The most important thing—and arguably the best thing—Robert
Graham's genius sperm bank did was to transform how Americans
thought about making babies.

Today, sperm banking is a business with "customers" instead of
"patients," marketing plans instead of doctor's orders, professional
donors instead of Johnny-on-the-spot medical students. None of this
was true when Robert Graham started the Repository. Sperm banking—and American fertility in general—experienced a revolution,
and Robert Graham was a most unlikely Thomas Jefferson.

To understand how sperm banks got the way they are today, you
have to start with a very short article in the April 1909 issue of the
journal *Medical World*. The article was titled "Artificial Impregnation." The author was an obscure Minnesota doctor named Addison
Davis Hard.

Hard wrote that twenty-five years earlier, he had witnessed a
medical procedure so bizarre and so shocking that everyone who
knew about it had taken an oath never to reveal what they saw. But
the time had come, Hard said, for him to break the oath. In 1884,
Hard continued, he had been a student of Dr. William Pancoast at
Jefferson Medical College in Philadelphia. Pancoast was a professor
of surgery, a former Civil War doctor who was just wrapping up a
fine though unremarkable academic career. A wealthy Philadelphia
businessman consulted Pancoast to learn why his wife wasn't getting pregnant. Pancoast examined the wife thoroughly—it being the

nineteenth-century (and twentieth-century, and twenty-first-century) assumption that if something was wrong it must be the woman's fault—and found nothing amiss. The husband, too, showed no obvious "physical defect," but Pancoast collected a semen sample just in case. He plugged it under the microscope: there were no sperm at all. Pancoast diagnosed the husband's sterility as the result of a youthful bout of gonorrhea. Pancoast was confident he could treat the problem and restore the husband's fertility. But when the businessman and his wife returned to Sansom Street Hospital two months later for the follow-up appointment, there was no change in the husband's condition. Pancoast told his students that the merchant was a lost cause: his seminal ducts were permanently blocked.

But then, eureka! According to Hard's account, "A joking remark by one of the class, 'The only solution of this problem is to call in the hired man,' was the probable incentive to the plan of action which followed."

Pancoast, without telling the husband or wife what he was going to do, knocked the woman unconscious with chloroform. The six students decided who was the "best-looking" man in the class, and the winner obligingly provided a sperm sample. Pancoast placed the semen in a hard rubber syringe and squeezed it into the unconscious woman's uterus. She woke up none the wiser. Later, Pancoast did deign to explain the insemination to the husband, who was "delighted with the idea," Hard reported. The husband and Pancoast then conspired to keep the secret from the wife, who soon found herself happily pregnant. This was the first recorded use of a sperm donor and probably the first use of a sperm donor, period.

In due course, a son was born. Remarkably, Hard said, he looked very like "the willing but impossible father." The boy, Hard added from the perspective of 1909, "is now a business man of the city of New York, and I have shaken hands with him within the past year." (Why would Hard, a Minnesota doctor, have any reason to meet a young New York businessman? As A. T. Gregoire and Robert

C. Mayer speculated sixty years later, Hard had no reason—unless of course *he* was that "best-looking member of the class," and thus the boy's father.)

Hard wrote that he was breaking his secrecy vow because Pancoast was dead and because he himself had concluded that donor insemination was a wonderful idea—a necessary eugenic remedy for a troubled nation. Too many men were secretly infected with gonorrhea, so babies needed protection from the "satanic germs" of their fathers. "Artificial impregnation" was the remedy. ("Go ask the blind children whose eyes were saturated with gonorrheal pus as they struggled thru the birth canal to emerge into this world of darkness to endure a living death; ask them" if they object to artificial impregnation, he wrote.)

It wouldn't matter which man supplied the seed, Hard concluded with the certainty of true ignorance, because the child was shaped entirely by the mother. The father made no contribution at all: "It may at first shock the delicate sensibilities of the sentimental who consider that the source of the seed indicated the true father, but when the scientific fact becomes known that the origin of the spermatozoa which generates the ovum is of no more importance than the personality of the finger which pulls the trigger of a gun, then objections will lose their forcefulness, and artificial impregnation will become recognized as a race-uplifting procedure."

I linger on the first donor insemination because it so perfectly prefigured the whole sorry history of sperm donation. From day one—from ejaculation one in Sansom Street Hospital in 1884—sperm donation has been characterized by its abundance of secrecy, embarrassment, deception, and ignorance. Hard's article encapsulated the business to come, all themes struck: the authoritarian fertility doctor making it up as he went along and issuing awesome decisions casually ("Call in the hired man") . . . the wife ignorant and unconscious, literally a mere vessel . . . the husband a coconspirator, lying to protect his own ego . . . the careless medical student, donating sperm without a thought to the consequences . . . the self-justifying scientific hoo-hah masquerading as empirical fact.

Artificial insemination is an elementary medical procedure—you can do it competently in your own home, after a few minutes' training—so why wasn't it practiced regularly until the twentieth century? One reason is that it took a long time for people to understand what sperm was or why it mattered, as Gina Maranto explains in *Quest for Perfection*. Until the seventeenth century, Maranto says, the male and female reproductive capacities were black boxes. Scientists knew about semen (obviously), but they couldn't figure out what the milky fluid had to do with reproduction. Some hypothesized that male semen mixed with a "female semen" in the uterus to form a child. Others guessed that semen fertilized menstrual blood. Finally, in 1677, Antoni van Leeuwenhoek, the Dutch inventor of the microscope, spotted sperm cells swimming in seminal fluid.

At first, sperm only confused matters. Another Dutch microscopist, Nicolaus Hartsoeker, speculated that a perfectly formed human being was curled inside the head of each sperm cell. These tiny men were dubbed "homunculi," and they became the foundation of a bizarre theory called "spermism." Spermism was a variation of the already popular notion of "preformation," which held that fully formed people already existed, microscopically, in their parents. Taken to its logical conclusion, preformation said that if your teeny son is housed in you, then your even teenier grandson is housed in him, and so on for N generations, till the end of time. We were God's own Russian nesting dolls. Until Hartsoeker, most preformationists had been matriarchal; these "ovists" believed that God had deposited us in the maternal eggs. (Therefore, Eve had held *all* of mankind to come, like some gigantic warehouse.) The homunculus flipped the attention to men, contending that humans came from sperm, and women were merely pots to plant in. The spermists and the ovists bickered for 150 years until they were both proven wrong. In the late eighteenth century, the Italian Lazzaro Spallanzani, performing what may have been the first artificial inseminations, showed that frogs and dogs couldn't get pregnant without contributions from both males and females. Then, in 1827, the mammalian egg was finally discovered. In the nineteenth century, European scientists ran experiments on animals, tinkering with sperm and eggs, eggs and sperm, and concluded, at long last, that each was useless without the other.

Meanwhile, doctors facing the practical problem of infertility wondered how to use sperm to help their patients. In the 1770s, the celebrated London physician John Hunter (my father is named after him, incidentally) arranged the first human artificial insemination. Hunter's patient suffered a penis defect that made it impossible for him to impregnate his wife, but he was still able to ejaculate. Hunter gave the man a syringe, told him to masturbate into the syringe's barrel, and then inject the semen into his wife's vagina. It worked. In the mid-nineteenth century, famed New York gynecologist J. Marion Sims used Hunter's method—along with some very painful surgery—to impregnate women suffering from "hyperesthetic" vaginas with their husbands' sperm. (The women supposedly had vaginas formed in such a way that they couldn't have intercourse with their husbands. "Hyperesthetic" is Greek for "Not tonight, dear.") Sims called his technique "ethereal copulation." The success rate was low—less than 5 percent of the women got pregnant, probably because menstrual cycles were poorly understood. Condemnation was emphatic. The Catholic Church denounced artificial insemination: be fruitful and multiply, yes, but not this way.

Hard's 1909 *Medical World* article was the first public hint that the new technique of artificial insemination could exclude the husband from reproduction. If artificial insemination using a husband's sperm was morally questionable, artificial insemination by donor (AID, as it came to be known) was anathema. Doctors were outraged by the mere thought of it. Some, with a striking ignorance of human physiology, insisted that what Hard described occurring at Sansom Street Hospital was literally impossible: a woman simply could not get pregnant in this way—certainly not without her husband's contribution. Others said that it was so immoral that it could not have happened. A doctor as noble as Pancoast would have been incapable of such a monstrous act.

But, immoral or not, AID was real, and it was useful, because it was the first effective fertility treatment. AID established the moral arc that all fertility treatments since—egg donation, in vitro fertilization, sex selection, surrogacy—have followed.

First, Denial: This is physically impossible.

Then Revulsion: This is an outrage against God and nature.

Then Silent Tolerance: You can do it, but please don't talk about it.

Finally, Popular Embrace: Do it, talk about it, brag about it. *You are having test-tube triplets carried by a surrogate? So am I!*

With AID, as with the subsequent fertility treatments, three potent forces combined to overwhelm the initial disapproval. First, the distress of the husbands and wives, who would risk anything to have a baby; second, the enthusiasm of doctors to try something new (and profit from it); and third, doctors' constitutional belief that *they*, not a backward society, should decide how their patients were treated.

After Hard's article, AID slowly progressed from the denial phase to revulsion. Then, in the 1930s, revulsion began to give way to silent tolerance. In 1934, Dr. Hermann Rohleder wrote *Test Tube Babies*, a history of artificial insemination and description of his AI techniques. He initially asserted that the only suitable purpose of artificial insemination was impregnating a wife with her husband's sperm and that donor insemination was outrageous: "What husband or wife, no matter how intense their longing for an heir, will consent to an injection of strange semen? Thank God that most people still have that much tact, decency, and moral feeling." Yet just a few pages later, writing as a doctor rather than a moralist, Rohleder conceded that he *would* impregnate a woman with a stranger's semen, under the right circumstances—if the husband was so desperate that suicide or divorce was a possibility, if the donor was healthy and unmarried, if the wife consented.

Rohleder's pragmatism would triumph, and silent tolerance followed for fifty years. The use of sperm donors spread slowly but steadily in the United States and Great Britain, the two pioneering countries. Starting in the 1930s, British doctor Margaret Jackson began discreetly providing freshly donated sperm to patients. Donor insemination took off in the United States after the War. In the Eisenhower era, doctors in the big cities began performing AID regularly. They collected sperm from colleagues, from medical students, and, dismayingly often, from themselves. By 1960, American doctors were creating 5,000 to 7,000 babies a year by donor insemination, up from essentially none a decade earlier.

AID was the only fertility treatment that actually worked, and parents were grateful for it. Still, it remained a secret and shameful ordeal. Most patients went to one doctor to get sperm, then another for the pregnancy and delivery, so that the doctor delivering the baby never knew that the father wasn't the biological father. Some doctors mixed sperm from the donor with the non-performing sperm of the father, so that the dad could pretend that his sperm had actually done the job. Doctors practicing AID usually kept no records at all, so there could be no chance of anyone finding out the truth. Doctors routinely signed false birth certificates, asserting that the sterile dad was the real one. The law encouraged such perjury: AID was technically adultery (still a crime in many states), and thus any child of AID was illegitimate.

In this first generation of AID, doctors tyrannized their patients. When a red-faced couple appeared at the office, mumbling about infertility, the doctor told them he would take care of everything. Mothers were discouraged from asking questions about the donor.

The doctor did a little poking around for a suitable donor—often the closest medical student at hand. The doctor would make sure the donor was the right skin color—white patients got white donors. If the doctor was feeling benevolent, he would also try to match the eye color of the father. (Oddly, another trait that doctors sometimes tried to match was religion, as though it had some genetic component.)

If the pregnancy took, doctors instructed parents—and husbands redoubled the instructions to their wives—that they were not to tell the child anything. They must pretend to everyone that little Jill was Daddy's girl, no matter how different father and daughter might look. The parents were told not to discuss it between themselves. They were even advised not to think about it. Fathers were ordered to behave toward the donor children exactly as they would behave toward their own biological children—advice that, not surprisingly, proved sadly impossible to follow.

This repression took its toll on families. Barry Stevens, conceived from AID more than fifty years ago, made a wonderful documentary, *Offspring*, about his search for his donor father. Stevens included clips of his own home movies, which show him, his mom, and his sister walking happily together and his sad-eyed dad trailing ten feet behind. His father, Stevens said in the movie, was always the family shadow—separate and unequal. The genetic disparity— mother connected by blood, father not—brought many DI kids closer to their mothers and drove fathers away. Most DI children never discover that their dad is not their dad. But those who do are rarely surprised; they always felt something wasn't right.

When I started writing about the Nobel sperm bank, my inbox clogged with e-mails from kids of the first big wave of AID. Now in their forties and fifties, many were sad and bitter. They told me the same story: *Dad wasn't like a real dad. When Dad died, Mom finally spilled the secret. Now I want to find my donor dad.* Their searches for their dads always fail. They just hit dead ends. They find that the doctor who inseminated their mom is dead. So are the nurses. The records have vanished or never existed. When they asked me for help, I always disappointed them. The best I could suggest was to advertise in the alumni magazine of the medical school where their mom got treated, because the donor might have been a student. But that has never worked either, as far as I have heard.

As AID became more common in the fifties, it started to poke its head out into the open, and society struggled with whether to welcome it or chase it back into its hole. Chasing was the first response. In the early fifties, a British parliamentary commission proposed criminalizing AID. The pope declared it a sin and recommended prison for doctors who performed it. In 1954, an Illinois state court ruled that AID—even with the husband's consent—was "contrary to public policy and good morals and constituted adultery on the mother's part." Thus, any DI child was illegitimate. (A 1959 British movie, *A Question of Adultery*, hinged on whether donor insemination counted as cheating on your husband.) But public policy gradually caught up with popular behavior. As American society loosened in the 1960s, attitudes toward sperm changed, too. In 1964, Georgia became the first state to legitimize DI kids. In 1968, the California Supreme Court held that a father who consented to AID for his wife couldn't later duck his paternal responsibility: he had become the child's legal father by signing his name to the AID contract. It did not matter that he had contributed no DNA. "Since there is no 'natural father,' we can only look for a lawful father," the court wrote. In 1973, the American Bar Association approved the Uniform Parentage Act, a model state law confirming that a husband is the legal father of a child conceived with AID.

New science also encouraged the spread of AID. In the first generation of AID, physicians relied on fresh semen, collected moments before from convenient interns and medical students. In 1949, a British researcher accidentally discovered that sperm frozen with glycerol could survive freezing and thawing. In the 1950s and '60s, American cryobiologists perfected the freezing process. They mixed the fresh sperm with a solution of glycerol, salt water, and egg yolk, then gradually cooled it down to minus 196 degrees centigrade in liquid nitrogen.

Fresh sperm had advantages: it was easy to handle, and it was very potent: women got pregnant from it pretty easily. But frozen sperm could be shipped. Even more important, frozen sperm meant you didn't have to rely on whatever donor was handy. Instead, you could stockpile the seed of all different kinds of men. You could have a bank.

Frozen sperm enabled AID to become a business. In the early 1970s, a few companies in big cities began collecting and freezing sperm in bulk and selling it to doctors who didn't want to wrangle donors themselves. From the start, sperm banking was a cowboy industry. The federal government didn't regulate it; neither did states. Anyone could open a sperm bank and usually did. Sperm banks were started not only by doctors but also by technicians, salesmen, and activists. Robert Graham—who called himself Dr. Graham— was an optometrist, as much a "doctor" and as much a fertility expert as I am. In most states, nothing stopped you from opening Fred's Sperm Bank and Delicatessen.*

---

* From the beginning, sperm banking had a comic aspect to it. In July 1976, a prankster named Joey Skaggs announced that he would be auctioning rock star sperm from his "Celebrity Sperm Bank" in Greenwich Village. "We'll have sperm from the likes of Mick Jagger, Bob Dylan, John Lennon, Paul McCartney, and vintage sperm from Jimi Hendrix," he declared. On the morning of the auction, Skaggs and his lawyer appeared to announce that the sperm had been kidnapped. They read a ransom note: "Caught you with your pants down. A sperm in the hand is worth a million in a Swiss bank. And that's what it will cost you. More to cum. [signed] Abbie." Hundreds of women called the nonexistent sperm bank asking if they could buy; radio and TV shows reported the aborted auction without realizing it had been a joke. And at the end of the year, Gloria Steinem—presumably unaware that it had been a hoax—appeared on an NBC special to give the Celebrity Sperm Bank an award for bad taste.

Amateurs went into sperm banking in part because banking and donor insemination were so easy. To open a bank, you needed a minimally equipped lab and some liquid-nitrogen tanks. And it was just as simple for customers. Doctors, trying to preserve their monopoly on insemination, had fostered the myth that AI was a complex procedure that only trained medical professionals could perform. In fact, it was a cinch. Anyone could do an insemination with a little training. You thawed the sperm, put it in a syringe attached to a "tomcat" catheter, threaded the catheter deep into the vagina (in some cases all the way into the uterus), and injected. Do-it-yourself inseminations—resulting in what were nicknamed "turkey baster babies"—had a brief vogue among feminists in the 1970s and '80s. Adrienne Ramm, the mother of three kids from the Nobel sperm bank, was inseminated by her husband at home. She said, "It was very important for us to make it a ritual at home, very important not to go to a doctor's office. It was a very mystical experience for my husband to plant that seed." (The Nobel sperm bank used to teach its clients the procedure. If a customer visited the Repository in Escondido, one of Graham's assistants might give her an impromptu lesson, right in the office, using a mirror and a few handy instruments—a Pederson vaginal speculum, a Makler insemination cannula, and a syringe. "So many women would tell me, 'That's the first time I have ever really seen myself,' " said Julianna McKillop, who directed the bank in the mid-1980s.)

Frozen sperm finally supplanted fresh sperm with the advent of AIDS. Some doctors who used fresh semen had collected from infected men; at least one woman contracted HIV from donor sperm. In the age of AIDS, the greatest advantage of frozen sperm turned out to be the delay it permitted between donation and insemination. Banks could test the donor when he gave the sample, store it for six months, then test the donor again to make sure he was still disease-free, thus ensuring that the frozen sperm was clean.

But even AIDS didn't prompt the government to pay attention to sperm banks. There were compelling political reasons why neither party wanted to start regulating fertility medicine. Lefties didn't want to tamper with sperm banking and fertility, because that would imply a government right to control what women could do with their own bodies. Abortion rights advocates feared that precedent. And the Right tended to ignore sperm banking and fertility because, although they were medicine, they looked like commerce. The free market was providing services that women wanted: Why mess it up?

In 1987 and 1988, at the urging of then-senator Albert Gore, the Office of Technology Assessment surveyed the American sperm industry—the only time then or since that the government has studied it. According to OTA's count, there were hundreds of sperm banks and more than 11,000 doctors performing inseminations. OTA estimated that 30,000 children per year were being born from anonymous donor sperm—which suggests that by now, there are about 1 million AID kids in the United States alone. OTA also found that only half of doctors kept *any* records of donor inseminations; and 2 percent of doctors admitted to having inseminated patients with their own semen.

Doctors could get away with inseminating patients with their own sperm, because fertility was still a field characterized by domineering physicians and timid patients. This became glaringly obvious in the "Sperminator" case of the late 1980s. Federal prosecutors indicted Dr. Cecil Jacobson of northern Virginia, aka "The Sperminator," on dozens of counts of fraud. Jacobson, one of the leading fertility specialists in the United States, was wildly popular with patients. But it turned out that he was giving infertile patients hormone therapy that made them register false positives on pregnancy tests. The women got their hopes up, only to discover they weren't really pregnant. He gave some women as many as ten false pregnancies. That was bad enough, but what really appalled the public was something else: Jacobson had also promised patients he would find them sperm donors who matched the characteristics they sought. Instead of doing that, he had simply inseminated them with his own sperm. Jacobson had fathered as many as seventy-five children this way. Patients were so awed by Jacobson that they didn't realize he was scamming them. (He was the very model of the authoritarian fertility doctor, prone to saying such endearing things as "God doesn't give you babies; I do.") Perhaps the most amazing fact about the Jacobson case was that inseminating seventy-five women with your own semen wasn't even a crime. In 1992, Jacobson was convicted on fifty-two counts, but self-propagation was not one of them. He surrendered his medical license and was sentenced to five years in prison. The fate of his children, whose identities were protected by the court, remains unknown.

The Sperminator case sparked enough interest in regulating sperm that the feds finally acted. In 1993, the Food and Drug Administration finally drafted regulations for sperm banks, requiring them to register with the FDA and screen donors for risk factors. But the first of those regulations did not take effect until 2004—eleven years later. Some of the regulations are *still* not in effect.

Robert Graham strolled into the world of dictatorial doctors and cowed patients and accidentally launched a revolution. The difference between Robert Graham and everyone else doing sperm banking in 1980 was that Robert Graham had built a $70 million company. He had sold eyeglasses, store to store. He had developed marketing plans, written ad copy, closed deals. So when he opened the Nobel Prize sperm bank in 1980, he listened to his customers. All he wanted to do was propagate genius. But he knew that his grand experiment would flop unless women *wanted* to shop with him. What made people buy at the supermarket? Brand names. Appealing advertising. Endorsements. What would make women buy at the sperm market? The very same things.

So Graham did what no one in the business had ever done: he marketed his men. Graham's catalog did for sperm what Sears, Roebuck did for housewares. His Repository catalog was very spare—just a few photocopied sheets and a cover page—but it thrilled his customers. Women who saw it realized, for the first time, that they had a genuine choice. Graham couldn't guarantee his product, of course, but he came close: he vouched that all donors were "men of outstanding accomplishment, fine appearance, sound health, and exceptional freedom from genetic impairment." (Graham put his men through so much testing and paperwork that it annoyed them: Nobel Prize winner Kary Mullis said he had rejected Graham's invitation because he'd thought that by the time he was done with the red tape, he wouldn't have any energy left to masturbate.)

It wasn't just that Graham offered choices, it was that he offered the *best*—the Godiva of sperm, prime cuts of American man. In Graham's catalog copy, the men were irresistible. He made them sound like men you could imagine talking to, men you could imagine taking a class from, and—above all—men you could imagine seducing. The physical descriptions included perfect, enticing details: "rosy cheeked, beautiful teeth." A donor's personality wasn't merely "happy," it was "happy and radiant." One of Graham's slyest marketing techniques was to scrawl handwritten comments on a catalog page— like throwing in the rustproofing for free: "Almost a superman!" he wrote on one.

Thanks to its attentiveness to consumers, the Repository up-ended the hierarchy of the fertility industry. Before the Repository, fertility doctors had ordered, women had accepted. Graham cut the doctors out of the loop and sold directly to the consumer. Graham disapproved of the women's movement and even banned unmarried women from using his bank, yet he became an inadvertent feminist pioneer. Women were entranced. Mother after mother said the same thing to me: she had picked the Repository because it was the only place that let her select what she wanted.

Where Graham went, other sperm banks—and the rest of the fertility industry—followed. California Cryobank, Xytex, Fairfax Cryobank, and the other major sperm banks started expanding their donor descriptions from a few lines to dozens of pages and recruiting the most gifted men they could find. Today, American sperm banks are heirs of the Repository for Germinal Choice, though they don't like to acknowledge its influence. To see the world Graham helped make, I paid a visit to Steve Broder at the California Cryobank, probably the world's leading sperm bank. Broder is a founding father of modern sperm banking. He was the technician who helped Graham collect sperm from Nobel Prize winners in the late 1970s. Later, he went on to cofound California Cryobank, where he was director of quality assurance. Broder was tall and lean. He was fiftyish and young-looking, except for gray hair brushed up in a style oddly reminiscent of Robert Graham's. Broder seemed good-natured but exacting, a characteristic you'd hope to find in a sperm bank's director of quality assurance.

Unlike most other sperm bankers, Broder acknowledges his debt to Graham. When the Nobel sperm bank opened in 1980, Broder said, it changed everything. "At the time, the California Cryobank had one line about a donor: height, weight, eye color, blood typing, ethnic group, college major. But when we saw what Graham was doing, how much information about the donor he put on a single page, we decided to do the same."

Other sperm banks, recognizing that they were in a consumer business, were soon publicizing their ultrahigh safety standards, rigorous testing of donors, and choice, choice, choice. This is the model that guides all sperm banks today.

Broder gave me a tour through California Cryobank, a low-slung redbrick building right next to the UCLA campus in Westwood. It also has offices in Palo Alto, right by Stanford, and in Cambridge, Massachusetts, within walking distance from MIT and Harvard. The locations are conscious strategy. California Cryobank realized that women wanted smart donors, so it went where the smartest boys were. Broder guided me through the back offices, past the five masturbatoriums. The door to one opened, and a slouchy, goateed twenty-year-old emerged holding a little cup.

We stopped at the lab, where the equipment and methods remained much the same as twenty years ago, Broder said. A white-coated technician invited me over to look into his microscope. He was counting sperm. His viewing field was divided up into a grid of squares. Each time he saw a spermatozoon in one of the squares, he clicked a counter. If he registered seventy-five live sperm in ten squares, the sample passed. That translated into a somewhat-above-average sperm count.

After the count, the techs mixed in cryoprotectant, the same formula of egg yolk, salt solution, and glycerol that's been used for decades. The sample was also processed to remove white blood cells, damaged sperm, and other material in semen that might interfere with insemination. Eventually, the mixture was pipetted into vials, each containing at least 10 million motile sperm. One ejaculation produced anywhere from three to thirty vials. The vials were threaded into metal straws and eventually transferred to the bank's liquid-nitrogen storage tanks, each of which holds 20,000 shots. More than six months later, the vials would be sold for $275 each. California Cryobank, like other major banks, uses only sperm collected on-site. The American Society for Reproductive Medicine and the American Association of Tissue Banks—the two trade groups that monitor sperm banks—strongly discourage allowing donors to collect and process their sperm at home. Processing the sperm in a lab improves its quality and ensures a reliable chain of custody. The bank can be sure the sperm came from the proper donor. (The Repository had permitted its donors to process their own sperm at home.*)

As the tech explained the storage process to me, a minidrama played out at the front desk. The slouchy, goateed donor was annoyed. The receptionist—a pretty young woman, natch—was gently admonishing him that his new sample didn't make the grade and that maybe he should try abstaining longer. "Well, do I get paid anyway?" Slouchy asked. "We can't pay you for a sample we can't use," she replied. "But I can give you free movie passes." She handed him a pair of tickets. The exchange felt like something out of a surrealist movie: Your masturbation is a failure. Here, have some movie tickets.

The consumer revolution in fertility has made the banks' job incredibly difficult. Imagine how picky you are when you shop for a CD player. Now suppose you expected the CD player to last eighty-seven years, occupy its own room in your house, get married, have children, and take care of you in your old age. You'd be pretty choosy, too. Sperm banks have to cater to that finickiness, or they fail. At the major banks, the attentiveness to consumer demand has reached extraordinary levels. Broder showed me how a prospective mom might shop for Donor Right at California Cryobank. He began by handing me the basic catalog, a three-page listing of all the cur-

---

* At home, even a careful donor could run into unexpected difficulties, as Donor Orange did. The Repository instructed home donors to first place samples in their freezer, to cool them down slowly, and transfer them to liquid nitrogen only later, after that pre-icing. Said Donor Orange:

"I was moving from one apartment to another down the hall, and I was in the middle of processing specimens. They were in the freezer of my old apartment. I wanted to make sure the electricity was hooked up in the new apartment so that the freezer would be working when I transferred the samples. I called the power company, and I didn't want to explain too much, so I told them that I had 'human specimens' in my freezer and wanted to make sure they were not damaged when I moved. The power company lady seemed taken aback, but she was very nice and confirmed that the power was on. I hung up.

"Ten minutes later the police were at my door. The officer wanted to come in and check the freezer to see that I didn't have body parts in it. I explained that the 'human specimens' were sperm donations. It was very embarrassing."

rent donors. The catalog is also online, which is where most customers view it. With two hundred–plus men available, California Cryobank probably has the world's largest selection. It dwarfs the Repository, which never had more than a dozen donors at once. California Cryobank produces more pregnancies in a single month than the Repository did in nineteen years. Other sperm banks range from 150-plus donors to only half a dozen.

In the basic catalog, donors are coded by ethnicity, blood type, hair color and texture, eye color, and college major or occupation. Searching for an Armenian international businessman? How about Mr. 3291? Or an Italian-French filmmaker, your own little Truffaullini? Try Mr. 5269.

But the basic catalog is just a start. For $12, you can see the "long profile" of any donor—his twenty-six-page handwritten application. Fifteen bucks more gets you the results of a psychological test called the Keirsey Temperament Sorter. Another $25 buys a baby photo. Yet another $25, and you can listen to an audio interview. Still more, and you can read the notes that Cryobank staff members took when they met the donor. For $50, a bank employee will even select the donor who looks most like your husband.

To get a sense of what this man-shopping feels like, I asked Broder if I could see a complete donor package. Broder gave me the entire folder for Donor 3498. I began with the baby photo. In it, 3498 was dark blond and cute, arms flung open to the world. At the bottom, where a parent would write, "Jimmy at his second birthday party," the Cryobank had printed, "3498." I leafed through 3498's handwritten application. His writing was fast and messy. He was twenty-six years old, of Spanish and English descent. His eyes were blue-gray, hair brown, blood B-positive. He was tall, of course. (California Cryobank rarely accepts anyone under five feet, nine inches tall.)

Donor 3498 had been a college philosophy major, with a 3.5 GPA, and he had earned a Master of Fine Arts graduate degree. He spoke basic Thai. "I was a national youth chess champion, and I have written a novel." His favorite food was pasta. He worked as a freelance journalist (I wondered if I knew him). He said his favorite color was black, wryly adding, "which I am told is technically not a color." He described himself as "highly self-motivated, obsessive about writing and learning and travel. . . . My greatest flaw is impatience." His life goal was to become a famous novelist. His SAT scores were 1270, but he noted that he got that score when he was only twelve years old, the only time he took the test. He suffered from hay fever; his dad had high blood pressure. Otherwise, the family had no serious health problems. Both parents were lawyers. His mom was "assertive," "controlling," and "optimistic"; his dad was "assertive" and "easygoing."

I checked 3498's Keirsey Temperament Sorter. He was classified as an "idealist" and a "Champion." Champions "see life as an exciting drama, pregnant with possibilities for both good and evil. . . . Fiercely individualistic, Champions strive toward a kind of personal authenticity. . . . Champions are positive exuberant people."

I played 3498's audio interview. He sounded serious, intense, extremely smart. I could hear that he clicked his lips together before every sentence. He clearly loved his sister—"a pretty amazing, vivacious woman"—but didn't think much of his younger brother, whom he dismissed as "less serious." He did indeed seem to be an idealist: "I'd like to be involved in the establishment of an alternative living community, one that is agriculturally oriented."

By then I felt I knew 3498, and that was the point. I knew more about him than I had known about most girls I dated in high school and college. I knew more about his health than I knew about my wife's or even my own. Unfortunately, I didn't really like him. His seriousness seemed oppressive: I disliked the way he put down his brother. He sounded rigid and chilly. If I were shopping for a husband, he wouldn't be it, and if I were shopping for a sperm donor, he wouldn't be it, either. And that was fine. I thought about it in economic terms: If I were a customer, I would have dropped only a hundred bucks on 3498, which is no more than a couple of cheap dates. I could go right back to the catalog and find someone better.

One of the implications of 3498's huge file—one that banks themselves hate to admit—is that *all* sperm banks have become eugenic sperm banks. When the Nobel Prize sperm bank disappeared, it left no void, because other banks have become as elitist as it ever was. Once the customer, not the doctor, started picking the donor, banks had to raise their standards, providing the most desirable men possible and imposing the most stringent health requirements.

The consumer revolution also changed sperm banking in ways that Robert Graham would have grumbled about. Graham limited his customers to wives, but married couples have less need to resort to donor sperm these days. Vasectomies are often reversible, and a treatment called ICSI can harvest a single sperm cell from the testes and use it to fertilize an in vitro egg. Now even a man who is shooting blanks has the chance to father his own children. The treatments are expensive, but they are gradually reducing the demand for donor sperm among married couples.

That means that lesbians and single mothers increasingly drive sperm banking. They now make up 40 percent of the customers at California Cryobank and 75 percent at some other banks. Their prevalence is altering how sperm banks treat confidentiality. Lesbians and single mothers can't deceive their children about their origins, so they don't. They tell their kids the truth. As a result, they're clamoring for ever more information about the donors to pass on to their kids. Increasingly, they are even demanding that sperm banks open their records so that children can learn the name of their donor. (Lesbians and single moms have also pioneered the practice of "known donors," in which they recruit a sperm provider from among their friends. The known donor, so nice in theory, can be a legal nightmare: known donors, unlike anonymous donors, don't automatically shed their paternal obligations. The state still considers them legal fathers. So mothers and donors have to write elaborate contracts to try to eliminate those rights.)

More and more married couples have also embraced the idea of openness. The old insistence on secrecy is out of fashion. Psychologists increasingly advise telling children about DI, lest the secret haunt and rend the family. (There is a small library of new books on how to tell children about their DI origins.)

This attitude change is prelude to a legal and legislative struggle. For decades, adoptees have been fighting a brutal, and sometimes successful, war to open adoption records. DI families are about to open a second front in that war. DI kids and moms argue that children have a basic genetic right to know their father's identity. Several countries, including Great Britain, Sweden, Switzerland, the Netherlands, and New Zealand, have recently ended donor anonymity and established donor registries for sperm and egg donors. When DI kids turn eighteen, they will be allowed to check the registries and learn their biological dad or mother's name. (The British registry starts this year, which means that the first kids who can search for their donors won't come of age until 2023.)

The United States is a long way from having a national donor registry, but DI families are beginning to organize and lobby for one. In the meantime, American sperm bankers are carefully watching what happens with the "identity release" program at the Sperm Bank of California. Way back in 1983, that progressive sperm bank inaugurated a program in which certain donors agreed that their children could contact them when they turned eighteen. The first crop of children have just reached majority, and a handful of them have now introduced themselves to their dads. Because these reunions are proceeding without too much trauma, other, bigger sperm banks are considering their own identity release programs. (This is a classic Europe-U.S. split. In Great Britain and Sweden, the government has imposed national donor registries to end anonymity. In the U.S., private sperm banks are trying to jerry-rig their own solutions to forestall government intervention.)

Some children, mothers, and donors are also circumventing the banks' anonymity policies on their own. The idea of using the Internet to connect sperm bank families has taken off in the past few years. Yahoo's Donor Sibling Registry has become a central warehouse where mothers, kids, and donors can advertise for one another. The registry, which is searchable by bank, mushroomed from a few dozen entries in 2001 to more than three thousand in 2004. These DSR listings read like singles ads, except that the object is a father or sibling, not a girlfriend.

> Hi my name is K—. I have blond hair and blue eyes. I'm looking for my father. He has blonde hair, green eyes and he is 5′9″.

There have been more than six hundred matches on the DSR so far. Most of these involve very young half siblings who are being connected by their parents. There are very few cases of children and donors finding each other, in part because so few donors have posted on the site. (Donor White is one of those who has posted.) For most American sperm donors, donating was something they did when they were quite young in order to make money. Most didn't spend a lot of time pondering the consequences of their action, because they didn't think there would be any. They counted on anonymity to shield them forever.

As anonymity crumbles in Europe and on the Internet, the sperm bank industry, now all too responsive to consumer sentiment, is being caught in a squeeze: it wants to deliver the openness its customers want, but it fears that the quality and quantity of donors will plummet if banks accept only men who are willing to be identified. In the Netherlands, for example, the supply of donors has dried up since donor anonymity was abolished. The American industry must choose between conflicting notions of consumer choice: the desire for a *known* donor versus the desire for the *best* donor. Customers, of course, expect both: incredible donors who are willing to be identified.

The consumer revolution in sperm banking has one more dark side that banks and families don't like to discuss. Once you start thinking of sperm as just another product, you start treating it like just another product. As long as customers consider AID to be a form of shopping, some of them will inevitably be disappointed. In America, the customer is always right. A customer who is unhappy with a DVD player can replace it. What about a customer who is unhappy with a child? Shopping for sperm looks like any other kind of commerce. There are products, marketing, competition. It's tempting to think that with enough knowledge, you will get exactly the child you want, as you can buy exactly the car you want. But sperm banks do make mistakes. They sometimes send out the wrong sperm. Sometimes they miss something dire in a donor's medical history.

More important, there is serendipity in DNA. A great donor can pass on a lousy set of genes. A recessive illness may be hiding somewhere, or just mediocrity. Women shop carefully for sperm in hopes of certainty. But there is no certainty in a baby. It does not come with a ten-year warranty. In sperm shopping, there is a deposit, but there are no returns, no refunds, no exchanges.

CHAPTER 10

WHO IS THE REAL DONOR CORAL?

```
                    DONOR CORAL #36                                    #)

SUMMARY:        A professional man of very high standing in his field
                has had a book published, excels in Mathematics.

ANCESTRY:       England, Norway

HAIR:           Thick brown

EYE COLOR:      Blue

SKIN:           Medium

HEIGHT:         5' 10"

WEIGHT:         165

APPEARANCE:     Very good looking

PERSONALITY:    Happy, slight extrovert, very easy going

BORN:           1950's

I.Q.:           160-at age 9

HOBBIES:        Writing, reading, chess and piano

ATHLETICS:      Excels in many sports

MANUAL DEXTERITY: Excellent

HEALTH:         Excellent

REMARKS:        Comes from a large family, all high achievers.
                Has 4 bright children with very good looks.

BLOOD TYPE:     O-
```

Donor Coral's entry in the 1985 Repository for Germinal Choice catalog.

A few weeks after Samantha Grant mailed her letter to the plastic surgeon Jeremy Taft, suggesting that he was Donor Coral, Dr. Taft sent her a reply. It was not what Samantha had expected. Taft denied that he was Alton's father, denied that he was Donor Coral, and denied, in fact, that he had ever donated sperm. Samantha was baffled and irritated. She wondered if he was lying. After all, she had first approached him more than a year earlier: If he wasn't Donor Coral, why had it taken him so long to deny it? And she noticed that he didn't sign the rejection letter? Why was that? (I suggested that he could be afraid that she had a sample of Donor Coral's handwriting to compare it to.) She reviewed all her evidence and was again struck by the coincidences. The similarities were so numerous: after matching age, personality, marital status, number and ages of children, hair and eye color, profession, geographic history, and hobbies, Samantha said, "you get down to a sample of one."

Half kidding, she mused about hiring a private detective to investigate Jeremy Taft and thought about how to obtain one of his fingernail clippings so she could DNA-test it. But then she allowed her disappointment to settle and disperse. She slowly let go of Jeremy Taft—not because she doubted he was Coral but because there was nothing to do about it. "I shall not bother him again," she told me—her formality made it sound definitive. If Jeremy Taft wasn't Coral, bothering him was pointless. And even if he *was* Coral, it was still pointless, because he clearly didn't want to meet his son.

This was in late 2002. Samantha and I rarely talked during the first few months of 2003. Then I made an unexpected discovery. There was one person who knew the identity of Donor Coral for sure: Julianna McKillop, who had managed the Repository in the mid-1980s and had accidentally given away Coral's first name when Samantha had visited the bank in 1985. When Samantha had started her hunt for Coral, she had tried and failed to locate Julianna, figuring that Julianna might be willing to call Coral for her. I, too, had searched for Julianna a little bit in 2001. But the letter I had sent her had been returned, and the phone number I had for her had been disconnected. I stopped looking for her; later, another donor told me she had died.

In 2002 and 2003, I was helping a Canadian documentary company make a film about the Nobel sperm bank. In spring 2003, the film's researcher, Derek Anderson, started looking for Julianna in the hope that she could find another donor Derek was looking for. Derek traced Julianna's movements—from the old address in California that I had for her to Europe and back again—and finally found her. She was very much alive, and living in San Francisco.

Derek gave me her phone number, and I called her. Julianna was delighted to be found. The Repository was the greatest adventure of her life, and she loved talking about it. We chatted for a bit about Robert Graham. Then, before I even mentioned why I was calling, she asked if I knew Donor Coral. He had been her favorite donor, she said. She had lost track of him when she'd left the Repository. She had always wondered what happened to him. "Oh, he was *wonderful*. I *loved* him. He was such a lovely man. Do you know what he's doing now? I'd love to see him again. We were *so* close. We were such great friends."

I said I was looking for Coral, too. A cat-and-mouse game ensued. She didn't want to reveal a confidence, so she wanted to make sure I wasn't lying to weasel information out of her. To convince her of my good intentions, I told her about Samantha and Alton, and Samantha's avid search. Julianna remembered Samantha, she was excited to aid Samantha's quest. Julianna asked me if I knew Coral's name. I said it was Jeremy, and that he was a doctor in Florida. That mollified her. She realized I was telling the truth. She asked if I knew Coral's last name. It was Taft, I said.

For a moment there was silence at her end of the line. She was considering whether to tell me the truth. Then she exclaimed, "That's not it!"

"It's not?"

"No, it's Jeremy *Sampson*, unless he changed it. Sampson is the name he always used with me. Yes, Jeremy Sampson, not Taft. Oh, how I would love to see him again. We must find him! How can we find him?"

Jeremy Sampson, *not* Jeremy Taft. The plastic surgeon hadn't been lying; he had been the victim of a very unlikely series of coincidences.

Julianna said she had always thought of herself as a "middle mom" when she was at the Repository—the apex of a triangular family consisting of the birth mother, the donor, and herself. She relished the prospect of middlemoming again. We agreed to search together for Dr. Jeremy Sampson in hope of reuniting him with Samantha, his son Alton, and Julianna.

I arranged to see Julianna in San Francisco and met her one Saturday morning in her beautiful downtown apartment. She managed the building and lived there rent-free. One of Julianna's many careers was as a painter, and she had decorated one wall of the dining room with a huge mural of her family. Practically every piece of furniture was painted in tempera—cheery, psychedelic patterns: I half expected her little dogs to be covered in swirls and paisley.

Julianna had gotten a raw deal from life but vigorously resisted self-pity: she had lost a daughter to cancer, one husband to a plane crash, another to illness. She was in her sixties, but carried herself with the energy of a younger woman; in the bathroom there was a recent photo of her skydiving. She had white hair and a too-deep tan, and a little Liz Taylor in her face. She reminded me, in fact, of a beloved, aging movie star, at once effusive and imperial.

I asked her how she had come to the Repository. "In 1981, I was a widow, I had just moved to La Jolla, and I went to a talk by [Graham's assistant] Paul Smith at the Unitarian Church in Del Mar. Paul was running the Repository at the time. I heard him and I thought, *That is the most fabulous idea I have ever heard.* I asked if I could work for him, and he said, 'Sure, I need someone to answer the phones.' " While Paul recruited donors and processed sperm, Julianna talked to the desperate customers who called asking for help.

Paul Smith and Julianna were a mismatched set. She thought he was slovenly and careless. His record keeping was not up to her standards, and he shed dog hair everywhere. ("I told Paul, 'Imagine if anyone gets infected because they are inserting dog hair in their uterus!' ") In 1984, she said, she convinced Robert Graham to fire Smith. Unfortunately for the Repository, the donors, who had all been recruited by Paul, left with him.

Paul Smith remained a fervent devotee of genius sperm banking after Graham sacked him. He took the Repository donors and opened a rival sperm bank, Heredity Choice, which is still going today. Heredity Choice has fathered nearly three hundred kids, according to Paul, more than the Repository ever did.

Today, Paul runs Heredity Choice out of love (or obsession)— God knows there is no rational reason to do it. He and his delightful wife, Adonna, own ten acres in the California desert, way out in the Antelope Valley, where they breed border collies, Siberian huskies, and genius babies. They make a teeny bit of money from selling sperm and puppies. They spend it on dog food and collecting sperm.

When I met him in 2001, Paul was still recovering from an embarrassing setback. The last time he had let a reporter come to his home was in 1996, when a *Primetime Live* camera crew filmed him for a story about genius sperm banks. Paul kept his Heredity Choice samples in a trailer that had no running water. That was unsanitary enough. But Paul also showed the crew his liquid-nitrogen tanks, where he stored human sperm and dog sperm side by side. Paul didn't understand why this was revolting. When ABC reporter Cynthia McFadden asked him about mingling the human and dog samples, Paul gave a funny, self-destructive answer: "The dog straws are twice as long [as the human ones]. I don't think I have ever confused the two. And none of my clients has ever had puppies from the sperm I have supplied."

California public health officials were understandably less jolly. They inspected Heredity Choice a few weeks after the camera crew's visit and ordered Paul to shut down—the only time California had ever closed a sperm bank. Paul squabbled with them a little bit, then moved his storage tanks to Nevada, which didn't regulate sperm banks.*

With Paul out of the way, Julianna convinced Graham to make her the manager instead. Julianna relished her new job. "I said, 'I am going to collect sperm come hell or high water.'" She cruised southern California in a white Pontiac Grand Am, searching for candidates. "I went to Caltech, and I started knocking on doors. I would say, 'I am Julianna, do you have fifteen minutes?' and he would say, 'Who are you?' And I would say, 'I am from the Repository for Germinal Choice. Have you heard of it?' And he would say, 'Come in and shut the door!'" When a donor accepted, Julianna would whip out a cup and tell him she'd return in forty-five minutes to collect it. When she got the sperm, she rushed it out to the car and, in the middle of the Caltech campus, mixed in the buffer and froze the vials.

Julianna shared Graham's conviction that improving the gene pool was the most urgent job in the world. When FedEx spilled a Repository liquid-nitrogen tank, thawing and killing the sperm inside, Julianna built a special wooden stand to hold the tanks upright during shipment. "One drop of spilled sperm was like *gold!*"

Eventually, she went so far for the cause that it got her fired. When a particularly desperate client failed to conceive with frozen

---

* The first time I interviewed Paul, he asked me if I could put him in touch with Bill Gates, since *Slate* was owned by Microsoft. Paul said he wanted to recruit Gates as a sperm donor—"even though I hate his operating system." I laughed him off. The second time I saw Paul, we were saying good-bye, and his wife, Adonna, asked, "If I send you a donor application, would you fill it out?"

"No, my wife wouldn't want that," I answered.

"C'mon, you'll think about it," urged Adonna.

"Okay, I'll think about it," I said.

As I drove away, my first thought was, *That is the most flattering thing anyone has ever said to me. I am genius sperm bank material.* I knew I would never do it, but how gratifying to be asked. But my second thought was: That is just sad. From Bill Gates to me—the perfect arc of decline.

Adonna, incidentally, never sent me the application.

sperm, Julianna suggested that her then boyfriend—a successful surgeon—contribute fresh semen, since fresh was more potent than frozen. The surgeon wasn't a donor of the Repository and hadn't been vetted by Graham and his board. And this was also *after* HIV had been identified. Julianna had been dating the surgeon for two years, so she vouched for his clean blood. That didn't protect her when Graham's wife, Marta Ve, heard about the proposed freelancing. Julianna was fired, though she said she would do the same thing again: helping a needy woman conceive was more important than an academic concern about a donor she knew to be clean and healthy.

The story of her firing brought Julianna back to her favorite subject and mine: Donor Coral. Julianna said that she never gave the client fresh sperm from her boyfriend. Instead, she tried once more with frozen sperm. That time she used a Coral sample. It worked, because Coral never failed. "He always had so much sperm, and it was so active," Julianna said. "And it is not just that there was so much of it, but they were all going in the right direction with one head and one tail. Jeremy's were like a whole school of sardines!

"Oh, Jeremy was wonderful. When Jeremy came along, that made the bank. We had almost no donors then, but he came, and he was so great, and I knew: now we can really have a sperm bank."

"How did you find him?" I asked.

He had volunteered, Julianna said. It was around 1984. "He read about it and he came to see us. I said, 'Tell us about yourself.' His answers were so wonderful." Jeremy wasn't a Nobelist. He was just a medical student, too young to have accomplished much. But he seemed just the kind of all-around stud that Graham craved—the type of man Graham's clients were so smitten with. He said he had an IQ of 160. He was gifted at chess and mathematics. He was a superb athlete. He came from a celebrated family of scientists and musicians. He was great-looking and incredibly charming. And he had already fathered three beautiful kids. "It was all so cool," said Julianna. The board quickly approved him as a donor.

Coral became the Repository's star. Julianna urged applicants to select him. She did that not only because she had a plentiful supply of his seed—he was an avid donor—but also because she thought, "His genes should go all over the place." Jeremy separated from his wife soon after he signed up for the bank, and Dr. Graham—keen to keep his prize stallion—told Julianna to make sure he was happy. She took him out to dinner on Graham's dime and bought him bottles of wine. They struck up a great friendship. "He was such a freethinker, so creative, so caring. He reminded me of Dr. Graham, in fact."

The Repository produced about sixty kids while she was there, she said, and half of them were Jeremy's. She had no doubt that Jeremy would want to meet his sons Alton and Tom. "Jeremy would want to see the whole family get together. He is a family man. He would love that."

I told Samantha about the discovery of Jeremy Sampson. She was overjoyed that the donor wasn't the objectionable Jeremy Taft. In mid-June 2003, she and I began searching in earnest for Jeremy Sampson. Julianna left the hunt to us. Julianna couldn't remember where Jeremy had worked or even where he had attended medical school. Still, we expected the search would be a cinch: How many Jeremy Sampsons had graduated from med school in California during the 1980s and then practiced in Florida? We examined Florida and California physician records, but there was no Jeremy, Jerry, Gerry, or Jeremiah Sampson. We found a Dr. Jeremiah Simpson, but he was way too old. I asked Julianna if she possibly misremembered the name. She doubted it, but we looked for other Jeremy Ss, anyway. I found a Jeremy Sanders who seemed promising, but Samantha saw a picture of him and knew it was the wrong guy. Samantha went to her local Mormon temple to check genealogy records. No Jeremy Sampson. She tried to cross-reference California's marriage and physician records. No Jeremy Sampson.

Samantha called medical schools in southern California. More dead ends. I proposed digging through California divorce records, since we knew Jeremy had split from his wife in the mid-1980s. Samantha considered hiring a research assistant to help us. We had mostly been searching online, but on a lark, I suggested that Samantha call the Medical Board of California and speak to a real person. Perhaps Jeremy Sampson had accidentally been dropped from the online directory.

Samantha phoned me the next day. Jackpot. The clerk at the medical board had found Jeremy Sampson immediately. She gave Samantha a work phone number for Dr. Sampson in Florida, at the State Department of Epidemiology. "It's great. It's great. It's great!" Samantha trilled to me on the phone. "This is amazing. This is him. It has got to be him. It's him. My God, the probability of this happening is so slim. . . ."

I asked her about the state agency he worked for. Samantha sounded a little dubious. "It doesn't sound like something someone of his accomplishments would be doing." Then she laughed at herself. "I have this dream, this idealized vision of what he is, and I know he can't possibly live up to it."

Julianna and I had agreed that she would make the first contact with Jeremy. We didn't want to spook him, which a call from me or Samantha surely would. I passed on Jeremy's phone number to Julianna. While we waited for her to call, I did a little more research on Jeremy Sampson. Now that I knew who the right guy was, it was easy to track him in online databases. I turned up some troubling stuff. He was involved in several lawsuits. He had had some minor run-ins with the authorities.

I didn't tell Samantha about most of this; I didn't want to alarm her. What little I did tell her bugged her, but not too much. She was eager to believe the best of him.

She told her son that Donor Coral had been found. Alton was pleased but cautious, Samantha said. He didn't know what to expect. He was curious about Jeremy's family history, but he definitely didn't want 'some kind of weird relationship.'" And he definitely didn't start talking about Jeremy Sampson as his "dad."

Samantha and Alton wondered how to approach Jeremy. They agreed that Alton should write a letter and send a picture. First, Alton agonized over how to address it: "Dear Coral" or "Dear Jeremy" or even "Dear Mr. Sampson."

Then he wondered what to write. "What the fuck do you say to someone like this? What do you say to him?" he asked his mom. They were both baffled. They were in a new world. There was no guidebook on how to meet your donor dad. "None of us has role models for this, it is unknown," Samantha told me. "We are on the edge of human feelings."

By now, Julianna had called Jeremy Sampson, and he had returned her phone message. He had been "ecstatic," Julianna reported, and wanted to know everything about Alton. He had offered to fly to Boston immediately, but Julianna had encouraged him to hold his horses. Julianna talked to Samantha and told her more about Jeremy. What Julianna said reassured Samantha. Jeremy and Alton were very similar, to go by Julianna's description. "He reminds me of my son in so many ways," Samantha told me after her conversation with Julianna. According to Julianna, Samantha said, "Jeremy is independent and creative. He does not care about what others think about him. He is often quiet and thoughtful and likes to assess a situation before he speaks." Samantha and I debated about whether we should also tell Tom that we had found his father. We decided to wait a little more. We thought we should let Jeremy adjust to one new son before springing a second son—and a grandson—on him.

In early July, a week after we found him, Jeremy called Samantha at home. They talked for almost two hours. Jeremy was fascinated with Alton. He asked lots of questions. The parallels between Alton and Jeremy's relatives flabbergasted Samantha. Alton played piano; Jeremy's mother had been a professional pianist. Alton was an aspiring marine biologist; Jeremy's father and grandfather were both celebrated marine biologists. Alton got on the phone for a few minutes, too. He and Jeremy compared likes and dislikes. Jeremy liked chess; so did Alton. Jeremy loved bike riding; Alton, too. They both preferred Russian composers to Germans. They exchanged photos by e-mail right afterward: they looked a lot alike, and Jeremy swore that Alton was a dead ringer for himself at sixteen. They made plans for Jeremy to fly up to Boston in August.

Samantha gave me the download the next day. "At the end of our telephone conversation, he referred to 'our boy.' I *loved* that. I never never never thought I would hear that expression."

Samantha was happy and stunned. Alton was dumbfounded. He told Samantha that it was "too much to grok." He wondered what in his life belonged just to him and what was programmed into his DNA. Marine biology—maybe that was a coincidence. But what about marine biology plus piano plus chess plus bikes plus Rachmaninoff? Alton posted cryptic notes on his blog. One began, "WhatamI?"

Samantha and Julianna kept me away from Jeremy. He didn't want to talk to a reporter yet. I wrote him a note through Samantha, assuring him that I wouldn't reveal his identity. Finally he wrote me back a letter and an e-mail. He said he was happy to talk to me. We made a phone date. From the moment he started speaking, I got curious vibes off Jeremy. He was undeniably sweet and friendly. He was admirably curious about his sperm bank kids. But his manner was vague and his conversation meandering.

I asked him about himself. He told me he had wanted a big family ever since he was a teenager. "Some people want fame and fortune. I wanted a lot of kids." Then he told me how many kids he had—just by his wives and girlfriends, not counting sperm banks: "about X kids with Y different women" was how he put it. To protect his identity, I can't reveal what X and Y were, but suffice it to say that X was an extremely high number, and Y wasn't small, either.

Jeremy had also been an avid sperm donor, he said. He had contributed to the Repository and two other banks. He was interested in meeting *all* of his sperm bank children. (In his letter to me, he had written, "Even a crocodile takes an interest in recognizing and protecting its offspring. Shouldn't a human being be interested in doing more than this?") He said he had stopped donating to sperm banks only because he had gotten distracted by real women. "I was involved with two women at the same time. There was not a lot of sperm left over."

I asked how he had managed to qualify for the Nobel sperm bank, since he had only been a young medical student at the time. "I was interested in Mensa. I had just broken up with my first wife, so I thought maybe I should follow up to try and get an intelligent wife or girlfriend. So I was reading about Mensa, and I must have seen something about the Nobel sperm bank. I was curious and I called them."

When the people there had interviewed him, he said, he had mentioned various distinguished ancestors and told them his IQ was 160. Had they asked for the IQ test results? No, he said.

"Is your IQ 160?" I asked.

"I don't know. I never took an IQ test. I told them the number I thought they would want to hear."

I knew the Repository had struggled for donors, but this was incredible. It had accepted a "genius" donor based on an invented IQ score.

Jeremy and I struck up a friendly e-mail correspondence. Despite his profligate breeding, I liked him. Partly I liked him because he was unembarrassed. He talked straightforwardly about the sexual aspect of donating sperm. He mentioned that he had even seduced women at sperm banks, recounting this disconcerting story:

"Once, when I arrived at a sperm bank, another sperm donor was arriving at the same time. He had a cast on one of his arms, so I said to myself, 'This guy is going to need some help getting sperm into a cup.' At that time, there were two attractive young women working at the sperm bank. I think they were college girls—premed students, most likely.

"For purposes of anonymity, let's call them 'Nancy' and 'Susan.' The four of us were standing there, so I said to the other sperm bank donor, 'I'll flip a coin, if it's heads, you get Nancy, and I get Susan to help me . . . if tails, I'll take Nancy and you get Susan.' Well, I flipped the coin, and I got Nancy. I took her with me into the examination room, where I was expected to masturbate to put sperm into a cup. I put my arms around her, and she blushed a beet red color and ran out of the examination room.

"I guess she was sort of innocent in matters of male sexuality. To this day, I don't know what happened between the other donor and Susan. Maybe the cast was not a real cast for a real arm injury but simply something that he wore so that he could get a cute female sperm bank employee to help him get sperm into the cup?

"Actually, at one of the sperm banks I did end up having a light-hearted sexual affair with one of the female employees that worked at the sperm bank. I say 'lighthearted' because although we were good friends, we were never madly in love with one another. It was more of a casual, sporadic, on-again/off-again sort of a relationship. Let's just say that she was very talented with her hands and I was able to donate sperm without my masturbating at all."

After her initial euphoria about Jeremy Sampson waned, Samantha began to worry about him. Jeremy had told her immediately about his many children, figuring she would find out about them eventually. She wanted to think that his greed to breed was harmless, but the more she learned, the more it bothered her. Jeremy warned her that if Alton ever wanted to marry another sperm bank child, they should make sure to get DNA tests. This disturbed her: Did he have that many DI kids, too? Some of Jeremy's relatives called Samantha to welcome her to the family, but also to tell her stories about him. They claimed he failed to support many of his kids financially. According to them, he had poor relationships with some of the children and no relationship with others. He was a reproductive opportunist, they said: he bred when he chose and left the parental responsibility to someone else. Samantha also learned that much of what she had been told about his family's accomplishments had been exaggerated.

Her correspondence with Jeremy soon grew sour and suspicious. Jeremy had been a gift, but already he felt like a curse. Every time Samantha learned something else unpleasant about Jeremy, he tried to brush it off, she thought. His imperturbability alarmed her. She wondered how he could be so unbothered by his chaotic life. Samantha and Alton started researching personality disorders on the Internet: Was there a condition that would cause someone to breed so indiscriminately?

I was asking myself the same question. Jeremy was the fourth Repository donor I had met who practiced this kind of reproductive excess. I had started to think of these guys as The Inseminators. All four had volunteered for the Repository, which wasn't surprising. If you have a compulsion to breed, of course you'd offer yourself to sperm banks. Two of the four—including Michael, the Nobelist's son—had just gone to sperm banks but had not fathered their own kids. This seemed egomaniacal—though not irresponsible—behavior. But two of them—Jeremy and another guy—had fathered both sperm bank kids and lots of their own. Jeremy and the other guy relied on the wives and girlfriends to do the work of raising their children. They seemed to believe that their genetic contribution was gift enough for the child. It occured to me that I might have stumbled on a new disorder: Onan meets Don Quixote meets Cheaper by the Dozen. I called some psychologists who specialized in sexual pathology to ask them if they had ever heard of men behaving this way. They hadn't and were intrigued. A couple of the psychologists characterized this compulsion to father children as an extreme form of narcissism. This kind of Darwinian self-involvement was a new phenomenon, they thought. Until recently, men were constrained in their breeding by the number of women they could seduce. No longer. Sperm banks allowed the Inseminators to reproduce without limit.

By mid-July, Samantha had begun to fear what would happen if Jeremy met Alton. Alton wasn't too keen on the idea, either. They decided to cancel the meeting. "My main concern is to protect my son," Samantha told me. When she recalled how Jeremy had said "our boy," it now infuriated her. Samantha wasn't brokenhearted. She was too angry to be brokenhearted. A so-called genius sperm bank, a four-year search, and *this guy* was the prize?

Samantha avoided Jeremy's e-mails for a few weeks. Finally she told him, "We are not going to see you."

Jeremy replied quickly, trying to provoke Samantha into changing her mind:

Dear Samantha:

The only information you have about me is hearsay from third parties, rumors, and innuendo (and perhaps some tall-tales or lies).

(Also, of course, you have my respectful and friendly e-mails to you, and we did speak once or twice on the phone, in a friendly and non-confrontational manner, if I remember correctly.)

What information are you using to base your decisions on? Did I say or do something to offend you or upset you?

When I was sixteen, my mother didn't try to tell me who to meet with and who not to meet with. She didn't tell me who to correspond with and who not to. . . . She never made any comments or suggestions about people that she didn't even know firsthand.

How long do you intend to "protect" Alton? Until he's eighteen? Until he's twenty-two? Until he's thirty?

After this e-mail, Jeremy proceeded as though nothing had happened, as though Samantha hadn't rebuffed him. He offered Alton the gift of a car, his old Saab. Jeremy told me he would visit Boston in August as planned. When I mentioned that to Samantha, she was incensed. She told him she and Alton would be away, no matter when he came. Samantha also told him to stop sending them e-mail, because she and Alton didn't want to hear from him anymore. Jeremy kept e-mailing anyway. Alton installed a block on his inbox. If Jeremy sent him mail, the autoblock bounced it with the reply "This message has been automatically deleted." Samantha put the block on her inbox, too. Samantha, whose e-mails had been all exclamation points and glee two weeks before, now signed her messages to me "Grrr."

Samantha told Julianna the reunion was off and that she and Alton didn't want to meet Jeremy. Julianna defended Jeremy. It didn't matter that he hadn't been a great dad, she said. "When I first heard about all his kids, I said, 'Oh darn. Why would he have all those kids?' But then I thought of Dr. Graham and how he used to always talk about brain drain. And I heard Dr. Graham talking in my head. 'Isn't this what we set out to do?' We *wanted* high-IQ children. We *wanted* babies born. *Whatever it takes.* 'The smarter you are, the more children you should have,' that's what Dr. Graham used to say." The kids were lucky to have Jeremy's genes, Julianna insisted. "If women are going to have these children anyway, isn't it better they do it with Jeremy? Isn't it better for them to have a high-IQ father?"

I worried that Samantha blamed me for the Jeremy debacle, so I flew to Boston to see her. She and I had still never met in person. I also wanted to clear up one thing that was nagging at me: Jeremy's accusation that Samantha was controlling her son and ordering him to avoid Jeremy. I wondered if Jeremy had hit on something. I wanted to see for myself.

Samantha collected me at the airport and drove me back to her house in Cambridge. She reminded me of a Sissy Spacek character: the sun-kissed farmer with an iron will who picks corn from sunup till past sundown, holds off a flood, and still looks good. She had beautiful skin. Her straight red-brown hair looked slightly archaic, as though she had gotten it styled in the nineteenth century. Her speech came in fits and starts. When she talked, it was a torrent. When she didn't, she could sit happily in silence for minutes at a time.

At the house, Samantha introduced me to Alton, who was wearing khaki cargo pants and a black T-shirt. He was lanky and cute, with thick hair, deep-set blue eyes, and a big cleft chin. By the time he hit college, he would be a catch. He greeted me politely but carefully. He shared his mother's watchfulness. Though he looked a bit like his half-brother Tom, they were opposites in manner. Tom's emotions and self-doubt were always visible. Alton was self-contained and exuded a quiet confidence.

We started talking about Jeremy. Samantha was both laughing and bitter. "I remember Julianna showed me his application, and I remember being really impressed with all his answers. None of it was true, of course." Almost nothing she had been told about Donor Coral actually applied to Jeremy Sampson. She pulled out his donor sheet and reviewed it.

"They said he was a 'professional man of very high standing.' *False*; he had just graduated from medical school—and not even a good one. They said he 'has had a book published.' Hah! The book was *self-published*. They said his IQ was 160. We know about that now. Julianna told me he excelled in math. Not true. She told me his sister had won international music competitions. *Not true*."

"And yet," she said, staring at her son, "it was the best decision I ever made." She grinned. "Despite Jeremy."

Alton didn't really want to talk about Jeremy or the Repository— he'd rather have discussed calculus—but he did it. He analyzed the

Repository and his origins unemotionally, as though standing next to himself.

"They say the Repository was wrong because it practiced 'selective breeding,' " he said. "I don't understand that. That is all we do, selective breeding. When you pick a wife or a husband, that is what you do, you get to know them and make sure you like them. That is selective breeding."

I asked him whether he wanted to meet Jeremy. I watched for any hesitation on his part, any sign that Jeremy was right and Samantha had made the decision for him. "I have no need to see the donor," he stated flatly. (Alton never uttered Jeremy's name while I was there; "the donor" or "the genetic donor"—that's what he always called him.) "I don't have some void in my life that needs to be filled. I have Daniel and my brothers." Daniel was his father, Samantha's ex. His "brothers" were Daniel's much older sons from an earlier marriage, who were not genetically related to Alton at all. "In a way we are closer because we are not related. They really *are* my brothers."

Are you *sure* you don't want to see Jeremy? I pestered. "I'm not interested. Maybe when I am much older I would meet him, just for curiosity's sake, but I would not be jumping for joy to do it. For now, I have no emotional need to do it. I just don't have an emotional gap that needs filling."

I was certain: this was Alton talking, not Samantha. He was his own self, the strong-minded son of a strong-minded mom.

It was clear that Alton was not like Tom or any of the other Nobel sperm bank kids I had talked to. I don't know exactly what genius is, but Alton was the smartest of the Repository by far. The house was crammed with evidence of Alton's accomplishments: stunning photographs he had taken, his piano, physics textbooks he had conquered, the schedule for his college classes, the iMac he made wiggle and shimmy. He shared his mom's analytical intensity and gift for explanation. Robert Graham and William Shockley would have recognized Alton as a kindred spirit: he broke every question—the Iraq war, the best route for a walk, a math problem, even his own conception and birth—into its component parts, then cracked it.

Graham would have congratulated himself on Alton's spark and declared it a tribute to Jeremy's glorious sperm. That would have been absolutely wrong. The most striking fact about Alton was just how much he resembled his mom. Their minds leapt and skipped in the same way. They emanated the same force field of reserved stillness. As far as I could tell, Alton had nothing in common with Jeremy but eyebrows, hair, and cleft chin.

The similarity of Samantha and Alton made me reconsider other Repository families I had met. In every case, the kids bore a remarkable resemblance to their moms. Tom Legare resembled his mother, Mary, in his striving, his juggling of work and family, and his emotional directness. In other families, too, the maternal resemblance was striking: the overachieving kids of overachieving Lorraine; an elegant, wispy, dancing daughter of elegant, wispy, professional ballerina . . .

The more I thought about it, the less surprising the maternal resemblance seemed. Most of these children had been raised only by their mothers. Their "social fathers" tended to be emotionally distant, and their biological donor fathers were out of the picture. So of course they were tied tightly to their moms. The mothers were women anxious for children, so motivated that they had chosen a genius sperm bank. Not surprisingly, they had become driven mothers. They spent more time with their kids than most parents did, certainly more than I did with mine or than my wonderful parents had with me. Was it any wonder their children grew up to be like them? I got the feeling that Samantha could have taken sperm from the dumbest player on the NFL's worst team and would still have raised a brilliant boy. Her good genes would have helped, but so would the stimulating world she created around her. Any child would have fallen under that spell.

And maybe there was also a genetic reason Alton was smart like his mom. Study after study has demonstrated the link between genes and what's called "general intelligence"—the ability to solve problems and think rationally. In aggregate, the more intelligent the parents, the more intelligent the child. It was this connection between genes and intelligence that made Graham sure that a genius sperm bank would improve humanity. But one emerging idea in genetics calls into question the value of genius sperm. It's called "imprinting." This is not the "imprinting" described by animal behaviorist Konrad Lorenz, which involves how animals bond to their parents. The imprinting I'm talking about concerns how genes are activated.

Here's the theory. A child carries two sets of every gene, one from each parent. Usually both genes are active, but some "imprinted" genes seem to be different. Only one of these genes is working: a signal tells the cell that only the maternal or paternal gene should be turned on. Cambridge University's Barry Keverne and Azim Surani have found that *maternally* imprinted genes (in mice, at least) are concentrated in the "executive" part of the brain—the areas that control high-level analytical thought and intelligence. *Paternally* imprinted genes, meanwhile, tend to be involved with the limbic system, which is the seat of emotions and primitive, instinctual behavior. In a discovery that will not surprise any big sister, Keverne and Surani found that mice created with only maternal genes had huge brains and scrawny bodies, while mice created with only paternal genes had scrawny brains and huge bodies.*

Imprinting is still a primitive theory, and no one would claim that Dad's DNA doesn't matter to his kid's intelligence. But imprinting does cast a shadow over Graham's grand plan. If the genius sperm

---

* Here is imprinting as an anecdote: The dancer Isadora Duncan once suggested to George Bernard Shaw that they should have a child. "Think of it!" Duncan said. "With my body and your brains, what a wonder it would be." Shaw replied, "But what if it had my body and your brains?"

is mostly contributing base emotions and big biceps rather than Quiz Bowl answers, who needs it? And imprinting makes Graham's indifference to the intelligence of his maternal applicants seem shortsighted. Maybe the mothers were the ones who mattered after all. Imprinting, in fact, calls into question the eugenic trend of the sperm bank business. Why recruit Phi Beta Kappans when jolly frat boys would do just as well?

On the other hand, imprinting may help explain the explosion of the egg donor industry—fertility's latest craze. Selling eggs has become a huge and seedy business as parents hunt for healthy, intelligent young women who'll surrender some eggs. Middle-aged couples—acting more like Darwinian auctioneers than aspiring parents—are trolling Ivy League campuses with ever-thicker wads of cash, placing ever more demanding advertisements in *The Harvard Crimson* and *The Stanford Daily*. An intelligent young coed can now collect ten, twenty, even fifty grand if she has healthy eggs to sell. Now add imprinting to the egg mania. If parents assume that maternal genes contribute extra to their children's intelligence, the egg bubble may get even worse.

After I read about imprinting, I sent Samantha some newspaper articles about it. She was delighted. It was a great relief to her. It meant that Jeremy was secondary—that his genes mattered less to Alton's intellectual development than hers. She forwarded the articles to him without a word of comment.

A few months later, she heard from him again. Jeremy, on a visit to Boston to see his sister, showed up unannounced at Samantha and Alton's house. He wanted to meet Alton. Samantha asked Alton if he wanted to come down to see Jeremy. Alton said no. Samantha asked Jeremy to go away, and he left without even seeing his biological son. By the end of 2003, Samantha and Alton had washed Jeremy out of their lives. Samantha and I talked less, and when we did, it was more often about her job, my job, or the war. The adventure had become a misadventure.

Samantha had lost the crusading enthusiasm she had when we first met. She had believed adamantly that DI children had a right to know their donor fathers. In our earliest phone conversations, she had grilled me about my attitude toward that question. I had explained that I thought kids did have a moral right to know but that you couldn't force donors to have a relationship with their offspring, merely to acknowledge them. This satisfied her, but again and again she had reminded me that we were on a crusade, not merely to find Donor Coral but to show the world the importance of children's genetic rights. The reality of Jeremy had deflated this.

"My thoughts have clarified as the result of the adventures of the past few months. We should not place great stock in meeting donors and half-siblings," she wrote me glumly. "One can develop a friendship with DI relatives, but a familial relationship is unlikely.

One of Jeremy's Repository sons, Alton, wanted nothing to do with him. But what about the other son? What about Tom? Unlike Alton, Tom had longed to find his donor dad. Did he still want to meet Donor Coral? Before Samantha's relationship with Jeremy went south, she and I agreed that it was time to tell Tom the truth. He had been kept in the dark long enough. I called him on his cell phone. I reached his voice mail. It was a new message:

> Hi, you have reached my answering machine, because I have amnesia and I feel stupid talking to people I don't remember. So at the sound of the beep, leave my name and please tell me something about myself.

I was not quick enough to leave the right message, which would have been "Tom, I have your name: It's 'Sampson.'" Instead I just said, "Call me. I have some news."

CHAPTER 11

DONOR WHITE FINDS HIS JOY

Donor White and Beth couldn't quite believe their good luck. The Father's Day e-mail from Beth unleashed a torrent of correspondence. Through the rest of June 2002, Beth and Roger wrote each other daily. They exchanged photos. The Repository had let Roger see pictures of Joy only when she was out of focus or too far away. For the first time since she had been a baby, Roger saw what his daughter really looked like. He thought she was beautiful. He bestowed his highest compliment: she was not merely pleasing, she was "highly pleasing."

Roger's e-mails to Beth stretched longer and longer. It all spilled out of him: his father's mathematical genius, his mother's poetry, his family's Revolutionary War history, his eighteen other White children. He apologized to Beth for his nearsightedness and hoped he hadn't passed it on to Joy. He asked every question he could think of about Joy. He was voracious; no detail of Joy's life was too trivial for his fatherly soul. He loved it when Beth told him how Joy had ridden twenty miles on a bicycle when she was six years old, how she had shone in last year's *Nutcracker*, how teachers loved having her in class, how she had written a book report about the ice skater Michelle Kwan.

Yet Roger found himself curiously ambivalent about actually meeting Joy. He felt embarrassed about himself—that maybe he wouldn't live up to Beth and Joy's expectations of him. He explained his recent illness to Beth, partly to reassure her that it wasn't genetic but also to warn her that he wasn't the same strapping man he'd used to be. He didn't want anyone to see him like he was now, maybe not even his lost daughter.

Ambivalence nagged Beth, too. As the correspondence rushed forward, Beth told Joy . . . nothing. Beth didn't let her know that her other father had been found. Eleven years ago, Beth had trusted Donor White enough to let him meet Joy. Seven years ago, she had trusted him enough to exchange letters with him. And she trusted him now, but, as they say in Washington, "Trust but verify." It was one thing to trust a man you would never meet, a man who didn't know your last name. It was something else to trust a man who hoped to be your child's (semi-)father. And it wasn't just Roger whom Beth had to worry about. Joy already had a dad and—since Beth had remarried—a stepdad. How many fathers could the girl handle? Beth waited a week, two weeks, three weeks. She read Roger's e-mails carefully, searching for warning signs. She saw none. This was a good man. Beth knew that he was not going to be Joy's father. She knew that Donor White would not try to supplant her ex-husband in Joy's heart—and wouldn't succeed if he did try. But she thought Donor White could be the grandfather Joy didn't have.

So on July 4, Beth told Joy about Donor White. The first thing Joy said was "I want to meet him!" When Beth mentioned that Roger had fathered eighteen other kids through the sperm bank, Joy celebrated: "Yay, I'm not an only child." Beth calmed her daughter. They couldn't meet immediately, Beth said. Roger lived in California, and he was in poor health. Still, that night Joy wrote an e-mail to Roger. She had nothing very profound to say. She told him how much she loved Harry Potter and ballet. She told him the names of her pets. She asked Roger what she should call him.

Roger was thrilled finally to hear from his daughter. *She's so well spoken,* he thought. He wrote right back. He suggested that Joy call him Roger. He promised to learn about Harry Potter and said the book he had loved when he was her age was *Lorna Doone.* A couple of days later, Joy e-mailed to say she had checked out *Lorna Doone* from the library. She instructed Roger to watch the first Harry Potter movie. If he liked it, then he must read the books. Roger's correspondence with Beth slowed, but he and Joy started trading messages two or three times a week.

A month later, Beth decided it was time for a visit. She still had friends near San Diego—holdovers from her California days—so she planned a visit with them at the end of August. Roger didn't have enough room in his house for them—his only guest room was stacked high with chemical journals—and Beth knew she wouldn't feel right staying there, at least not right away. I begged Beth and Roger to let me tag along on the visit. I tried guilt-tripping them ("You never would have met without me"), to no avail. They insisted this family reunion (or was it a "union" rather than a "*re*union"?) be private. I didn't blame them. Still, I was disappointed. They were making history. As far as I could tell, this was the first time an American child would meet her anonymous donor father.

Beth and Joy showed up at Roger and Rebecca's house at 9:30 A.M. on Sunday, August 18. There were no blinding lights, burning bushes, or gasps of amazement when father met daughter. Joy hugged Roger as soon as she saw him. She gave him a T-shirt, a prize from a recent soccer tournament. He offered her two T-shirts in return, souvenirs from 10K races. He worried that they were too large. Don't worry, Joy said, I'll use them as nightgowns. When Roger and his daughter talked, his wife, Rebecca, took Beth aside and thanked her for bringing Joy. Rebecca's thanks was an act of grace that won

Beth's heart. After all, Beth and Joy had flown across the country to thank *them*.

The four of them spent four happy days together—"perfect" days, Beth said. All the adults shared the same interest: Joy. Roger and Rebecca wanted to watch her and listen to her, Beth was happy to talk about her, and Joy was blithe enough to go along with all of it. Beth and Joy had toted a suitcase full of photo albums and video-tapes across the country, and Roger and Rebecca passed many happy hours watching Joy grow from infant to adolescent. They watched the tapes of Joy's ballet recitals, oohing and aahing at the right moments. Roger showed Joy his family scrapbooks, as well as the little black book where he commemorated his Repository kids. Roger unfolded a family tree and walked Joy through it all the way back to their Pilgrim ancestors. She delighted him by asking tons of questions about it: Where did that guy get his name? How is she re-lated to us?

They strolled on the beach, where Joy turned cartwheels, leapt over piles of kelp, and bodysurfed. She collected seashells for a keep-sake picture frame. At night, they burned a candle that Joy had made. And they snapped endless pictures. I saw the photos a few months later. Joy, in sunglasses and shorts, all tan limbs and teeth, practically bounced out of the frame—jumping, mugging, hugging. Roger, the object of her hugs, looked stunned, as though he couldn't believe he had received such a present.

A few days after Joy left, Roger e-mailed me: "Our home feels more lonely and empty now." Rebecca had been anxious about the visit, Roger said, but Joy and Beth had won her over, too. Roger summed up Joy. He was giddy but still meticulous:

> I was not able to find one thing about her that I would wish to change, even down to the smallest of details. Let me at-tempt to list what I liked best about Joy, in order of impor-

tance: (1) she is healthy, happy, well mannered, modest, un-
spoiled, and is considerate of others; (2) she has obvious
talent in dance and music; (3) she is athletic and does well
in several sports; (4) she has a sharp and quick mind that
allows her to take new facts and rapidly use them to make
interpretations that are truly surprising for one only 12
years of age; and (5) her appearance is very pleasing. Indeed, I believe that most people would agree with me that
she is beautiful, but Joy herself says that appearances are
unimportant and that it is the quality of the person within
that really matters. Needless to say, she has captured our
hearts forever.

Roger was besotted with his daughter. Everything she did was
stupendous, everything she wrote was brilliant, every dance was perfect. When I slipped up and referred to her as a thirteen-year-old,
Roger rebuked me: "Why David, she is only twelve! Don't you know
she won't be thirteen till September?" Everyone, he seemed to believe, must know everything about Joy. She deserved it. Once when
he told me a little story about something she had done—a small,
sweet favor—he was so happy that he wept. Roger was amazingly,
sometimes awkwardly, grateful to me. He called me "an agent of
destiny."

Roger started referring to Joy as "my daughter." But he was sensible enough to know that she didn't feel the same overwhelming
love for him. She was his daughter, but he was not her father. He accepted that. "I realize that her feelings toward me are more diluted
by the fact that she has a father and a step-father, whom she cares
about very much. This is perfectly acceptable to me, because Joy has
enough room in her heart to find sufficient affection for me, I believe."

During the months after the visit, Roger emerged from the cof-
fin where he had buried himself during his illness. His sickness had

made life feel pointless. He had thought he had squandered his good years on work and was fading into a useless, incapacitated old age. Joy revived him. He started consulting again for the chemistry companies he used to work for—surprised and gratified that his expertise was still in demand. He wrote magazine articles about his hobby, Civil War history, and gathered string for a biography of his favorite general, Stonewall Jackson. (He even pitched me an article for *Slate*, though we didn't publish it.) Roger also became a kind of fairy godfather of sperm banking. He visited the online sperm bank chat sites, particularly the Donor Sibling Registry forum, where children and sperm donors hunt for each other. He recounted his happy reunion with Joy, hoping to inspire other kids and donors. It could happen to them, too.

Roger and Joy kept e-mailing a couple times a week, easy little messages, nothing deep. He reveled in the chitterchatter. He gravely considered her plans to plant tulips in the garden. Was her new "invisible" retainer really invisible? he asked. Next time she visited, would he even be able to see it? When she wrote him a few sentences, he responded with a few pages. He couldn't help it. His happiest moments were when Beth asked him for advice about Joy—everyday things, such as, should she quit soccer to have more time for ballet? He loved that they thought his opinion mattered—or even that they *pretended* that it mattered.

Finding Joy made him both more and less curious about the other eighteen White kids. Sometimes he really wanted to search for others, but then he checked himself. What if they weren't marvelous like Joy? "Joy is a home run with the bases loaded," he told Beth. But maybe the others were broken-bat singles or even pop-outs. Roger did consider following up with the son he knew about—Jeroboam, the boy who had once greeted him, "Hello, Man in the Hat," as he jogged by. Roger learned that Jeroboam was now at the local high school, where he ran track. Roger thought he might go watch his son at a track meet. Then he'd figure out what to do.

A few months after Beth and Joy returned from their visit with Roger, I drove to Pennsylvania to see them. I had only talked to Beth on the phone, never met her. I was curious if finding Roger had been a life-changing experience for them, too.

They lived in a small rural town—not a hick town, it had a college. Beth and Joy met me in front of the Dairy Queen, and we drove over to her house on the outskirts of town, where they lived with her husband. There was a big backyard, and countless dogs were underfoot. The house was a welcoming kind of house—a house, in fact, with a "Welcome" sampler mounted in the front hall. Joy's soccer trophies and beauty pageant ribbons filled a glass case in the living room.

Beth was tall and pretty, in a serene way. She had long blond hair, wonderful teeth, and great posture. Joy looked like her mom, but with more mischief in her face. She was built like a dancer, with long limbs and hardly an ounce of fat. She was ebullient, where her mom was quiet. Joy seemed a little younger than the thirteen-year-olds I knew in Washington. Her small-town childhood may have sheltered her from cynicism and low-rider jeans.

Beth was clearly proud of Joy, but in an understated way. "She's a good kid" was how she put it. Beth showed off Joy, but only because I asked her to. I spotted Joy's report card pinned to the fridge. "There's not much room for improvement." Beth beamed, pointing to the unbroken line of As. I commented on the harpsichord in the living room—an unusual instrument for a thirteen-year-old. Beth asked Joy to play for us—Pachelbel's *Canon*, and several variations on Cat Stevens's "Morning Has Broken," which Joy interrupted with "Drat" whenever she dropped a note. Joy gave me a tour of her room. It was all girl: canopy bed, wallpaper with a lily pattern, the requisite Harry Potter poster on the door, a shelf full of keepsakes.

Joy asked if I wanted to see her ballet recital, so we ate take-out pizza in the basement and watched last winter's *Nutcracker*. Joy gave a play-by-play on which girls were good and which were her friends and which part she hoped to dance next year. All the while she raced around the basement, doing twirls and leaps and splits.

Joy was plenty smart, but was she a genius or superkid? I doubted it. Beth didn't think so. Just as I saw much of Samantha in Alton, I saw much of Beth in Joy—their all-Americanness, their good nature, their grace. Joy wasn't an Einstein, but she was the kind of girl that teachers want in their class, the kind of girl you want your own daughter to be friends with, the kind of girl you want your son to take to the prom.

Joy loved talking about Roger. (That's what she called him, not "Dad.") When I first asked her about him, Joy sprinted down the hall to Beth's bedroom and raced back carrying a careworn Playskool doll. "That's the doll he gave me." Joy showed off a lily pendant that Roger had sent her, which she never took off. (Lily was her middle name.) She pulled out her latest project: she was embroidering a sampler for Roger as a Father's Day gift. Both mother and daughter were proud of Roger's professional accomplishments, too. They handed me a photo of Joy sitting on Roger's couch, cradling in her arms the three books he had written—all very abstruse technical manuals. "I don't understand *any* of what he does," said Beth, full of admiration.

Beth chided me for still calling Roger a "sperm donor." When he had held Joy as a baby, she said, he had "identified her as *his* in some strange and wonderful way." Now that he had gotten a chance to meet Joy again and know her, he was not a sperm donor anymore, she insisted. He was a *dad*.

But not really Joy's dad. That was a more complicated question. Joy's stepfather was out of the house when I came; Beth told me that he thought they should be careful around me, and he didn't want any part in the story. And Beth would not tell me anything about Joy's father, would not let me speak to him, and discouraged me from talking to Joy about him. His relationship with Joy, she implied, was strong and close, and it was none of my business. So I didn't know whether the discovery of Donor White had altered Joy's relationship with her father or not.

Joy was at the ideal age for meeting Roger, in the oasis right before adolescence. The hormones of fourteen, the brutality of fifteen, the rage of seventeen were all awaiting her. Now she was still a girl. Meeting Roger had been a joyful adventure, but it hadn't changed her conception of herself. She was too young, and too mentally healthy, for it to shake her identity. She was more interested in who would dance the Sugar Plum Fairy next Christmas than in who had given what sperm to whom. I was not sure that a thirteen-year-old could even really understand who Roger was and what he had done. Of course, she might comprehend it in the simplest sex-ed, making-babies sense. But how could a thirteen-year-old process the idea of a fifty-six-year-old man collecting ejaculated semen from a condom, freezing it in liquid nitrogen, and giving it to a reclusive eugenics-obsessed millionaire who had then FedExed it to women all over the country, including her mother? What box in a thirteen-year-old's brain would that fit into?

Joy's feelings for Roger did not seem complex: She admired him. She liked him. She even loved him. But she was not asking herself why he had done what he had or how much of him was in her and what that meant. Joy's life was not so unsettled that Roger filled a void. Like Alton, she had no void. Her father was still her father. She did not need to be completed.

Beth told me a story from their visit with Roger. When the time had come to leave, Beth said, Joy had had a hard time saying good-bye to Roger and Rebecca. She had been choked up. But half an hour later, she had been playing happily with friends. "I told her that daily life was not going to change after meeting Roger," Beth said. And it hadn't. Roger had a new daughter. Roger had a new life. All Joy had—and it was plenty—was one more person to love.

CHAPTER 12
A FATHER, FOR BETTER OR WORSE

I was waiting for Tom to call me back so I could tell him about his father, Jeremy Sampson. I felt grim. Tom had begun his search expecting to find a Nobel dad, a genius, maybe even Jonas Salk. Instead, he was getting Jeremy—an obscure doctor whose notable accomplishment in life was leaving a wake of ex-wives and forgotten children. There's nothing worse than a wish unfulfilled, except a wish fulfilled.

So I wanted to try to make it gentle for Tom. When he returned my message, I told him the big news. I said that Samantha and I had found his father, his name was Jeremy Sampson, and he was a doctor in Florida. I didn't mention all Jeremy's kids, his erratic past, his made-up IQ, or his exaggerated accomplishments. Tom, who was never speechless, was speechless. After a little bit, he managed to mutter, "Wow, great" and "Thank you" and "I can't believe this." I told him to expect a call from Jeremy the next day. Before I hung up, I suggested, very shyly, that perhaps Jeremy wasn't exactly the ideal father and that perhaps Tom shouldn't expect Jeremy to surpass all his hopes and dreams. Tom was too shell-shocked from the headline—Dad Is Found—to listen to my caution. I hoped for the best.

Tom felt more excited than he had for years, but he warned himself to calm down. He knew he was an easy mark. He was too ready to believe the best in others, and he had been burned repeatedly because of it. Only when he had met Lana, who was full of Slavic pessimism, had he realized that he lacked common sense and critical judgment. Now he was aware of his naiveté, and he tried to order himself to be careful and not to hope for too much. But he couldn't help it. His parents' divorce had just been finalized. Maybe discovering Jeremy now was more than a coincidence. Just as he was losing his old father, here came a new one to take his place. Perhaps Jeremy would be the dad his dad had never been.

Back in Florida, Jeremy was happy enough to learn about Tom—another notch in his Darwinian bedpost—but was genuinely thrilled to hear that Tom had a son of his own. "I am amazingly happy elated shocked and surprised that I am a grandfather!" he told me. I gave Tom's phone number to Jeremy and told him when he should call.

The day arrived. Tom took off from work. He spent the afternoon pacing around the house, playing video games, and staring at the phone. Sometimes it rang, but it was always a friend or someone calling for his mom. At 8:30 P.M., it rang again, and Tom knew this was the call. *That's my dad calling.* My dad, he thought. He answered it on the first ring.

"Hello, this is Jeremy. I'm calling for Tom."

Tom had rehearsed this moment over and over for the past two years. In the conversation he had imagined, he would be angry. The rage would spill out of him. He would confront Donor Coral, grill him on why he had thought it was okay to jack off and leave.

But instead his mind went empty. All he could think to say was "Hello." And then "So you're my dad."

Jeremy said, "Yes."

The conversation scraped along with chitchat. Jeremy tried hard. He asked for Tom's address. He asked about Darian. He asked about the weather in Kansas City. For Tom, it all felt out of body: *I am talking to my dad, and I have nothing to say.* Tom thought Jeremy sounded old. Still, Jeremy's curiosity comforted Tom, because he had feared that Jeremy would be chilly and distant. Jeremy called Tom "Tom." But Tom didn't know what to call Jeremy. He still couldn't decide. Finally, after a few minutes, Tom tried saying "Jeremy." It felt okay. Secretly, just a little bit, Tom still thought of him as "my dad." Tom invited Jeremy to visit him in Kansas City, and Jeremy reciprocated by inviting Tom to visit him in Florida.

The conversation flagged after ten minutes. Jeremy's eight-year-old daughter, Stacy, pestered her dad to hand her the phone. She didn't know who Tom was, but they talked for a little bit about nothing. Then, at Tom's end, Darian hit his head on the floor and started screaming. Tom had to comfort him, so he said good-bye to Jeremy and that was that.

I called Tom the next day to find out how the conversation had gone. He told me, "It was great," and said he was really happy. Then he confessed how stilted the conversation had been. He said he already feared Jeremy was going to disappoint him, that they were not going to have a real relationship.

A few days later, Tom got a call from Jeremy's sister. She said she wanted to welcome Tom to the family. She filled Tom in on the Sampsons' history. She told him he might be a great-great-great-great-great-grandson of Ben Franklin. He might also be a descendant of Nathaniel Hawthorne and of one of the Salem witches. The sister also boasted about how brilliant Jeremy was. She insisted he was a genius, though Tom was dubious. Jeremy had sounded too much like a regular guy on the phone. Then she delivered a warning. She told Tom about Jeremy's many, many children and his spotty treatment of them. Tom was stunned by the news but he was determined not to judge Jeremy till he knew him better.

During the next couple of months, Tom and Jeremy squeezed in only two short phone conversations, both abbreviated when Darian started to cry. Both calls frustrated Tom. He and Jeremy weren't getting closer. Tom yearned for a visit. He wanted to show off Lana and Darian to his donor dad. It soon became clear that Jeremy wasn't coming to Kansas City, so Tom arranged a visit to Florida. Tom, Lana, Darian, and I would fly to Miami in early September and stay with Jeremy for a couple of days. We would also meet Jeremy's latest female companion—possibly a wife, he called her indeterminately his "old lady"—and their two daughters.

Tom, Darian, Lana, and I rendezvoused at the Miami airport at midnight on a Friday. It was Darian's first plane ride and Tom's first trip to the East Coast. As they waited outside the terminal for me to pick them up, Tom and Lana stuck out in the colorful Miami crowd. They were wearing their regular uniform: black Insane Clown Posse T-shirts with glaring satanic clowns on the front. Darian was in his baby seat, cheerier and more active than when I had seen him six months earlier. He had a shock of blond hair and a permanent wide-mouthed grin, like a happy hippo.

Tom's voice was softer and mumblier than usual. He sounded nervous. "I have that night-before-Christmas feeling," he said as we drove to the motel. "I'm scared and happy and excited."

Tom was worried about juggling his two dads. He feared he was trading Alvin, a dad who was imperfect but familiar, for the unknown Jeremy. "I told my dad—my first dad—I was coming out here. My dad really wasn't too happy about it. He didn't say much. But he told me he didn't want me to go."

At the motel, Tom told me he was frightened that what was inside Jeremy—the compulsion that had made him sire X children and not take care of them—was also inside him. "The dad who raised me was not a good dad. I was really hoping Jeremy would be a good dad. I am already scared I am going to be a bad dad to Darian."

Alvin had taught Tom nothing about being a good father. If

Jeremy was also a bad father, that meant that Tom's genes were stacked against him, too. So nurture *and* nature were conspiring against him, directing him toward paternal incompetence, indifference. I tried to reassure Tom that his two dads didn't matter. He was already proving himself a good dad, I said. After all, I pointed out, at the very moment he was expressing this fear, Tom was hunched over a motel bed, dabbing the spit-up off Darian's yellow onesie. His family would be fine, I told him, if Tom trusted himself and not his DNA.

Discovering you are a genius sperm bank kid can muddle you in all sorts of ways, but the worst may be in causing you to suddenly believe in genes. Before Tom discovered he was a Nobel sperm bank baby, he had never thought about whether his DNA had made him the way he was. There had been no reason to. But once he learned that he had a special genetic heritage, he applied genetic thinking to his whole life. If he did something well or badly, he would credit it to the genes. When he thought about his future, he tried to read his DNA like a palmist reads hands: *What does the double helix say I should do?* And now that he had found Jeremy and learned his dubious history, Tom was letting the genetic perspective rule him again. He was discounting the evidence of his very own life—the fact that he was working his eighteen-year-old butt off to raise, nurture, and financially support his baby son—because his genes seemed to contradict it.

Tom was doubting more than just his parenting skills. He seemed uncertain about everything, as I saw the next morning, when I invited him and Lana to accompany me while I visited a friend who worked at a Miami radio station. Tom was mesmerized by the station's engineering booth. He left the studio talking a mile a minute about how he should be an audio engineer. A few minutes later, when we stopped at Subway for breakfast, he reconsidered. He said he didn't know where he could get trained.

This segued into a tentative, hesitating monologue about his career plans. "For the past four months I've been racking my brains about what I want to do. Sometimes I think I want to be a lawyer or a veterinarian. Or maybe I want to be a doctor, but maybe I can't. I don't know. Sometimes I think I want to write novels, but that is really hard to get into. Or I could just stay at my mom's company, I guess." I asked what he really loved to do. "I really like to play video games—that's what I really love. But there are no jobs, except being a game tester, and that's even harder than getting a job writing novels."

After breakfast, we started driving out toward the distant suburb where Jeremy lived. It was a clear day but viciously hot. Tom called Jeremy from the road to let him know we were on the way. When he hung up, Tom was back in good spirits. "Jeremy kept asking me about Darian, checking to make sure we have the AC on in the car. He also wanted to know if we are all going to stay with him—he says we'll have to squeeze, but that it will be fine. It's weird, I haven't even met him, but it seems like he really cares about us. Whereas my dad who I have known for eighteen years doesn't care about anyone."

Then, out of the blue, Tom announced that he and Lana were married. They had sneaked off to a justice of the peace two weeks before. There had been no witnesses and no party. They had told no one before they went and no one after. Tom's mom had figured it out a few days later, when she had noticed the marriage license poking out of the diaper bag. She was pleased, because she had feared that Lana could kidnap Darian back to Russia if they weren't married. Mary had been trying all kinds of different arguments on Tom to get him to tie the knot. The one that had worked was money: Tom's car insurance would drop by $115 a month if he got married. That's why they had done it, Tom said. ("Yes, not for love," Lana injected dryly.) Tom did not say why they had married so suddenly, right before this trip, but he didn't have to. Tom was eager to present Lana to Jeremy as his bride. Tom mentioned that he and Lana had even considered flying to Miami a few days early and getting married with Jeremy there.

As we drove, Tom sank into a Jeremy-focused reverie, his mood shifting in almost every sentence. "I have a feeling I may end up without a good relationship with either Jeremy or my own dad. It won't be a real father-son relationship with Jeremy, that's what I am worried about. I can talk to my mom about anything. She knows how my life is going, and I know how her life is going. She tells me that she loves me, and she tells me that she cares. That's the kind of relationship I would want with my dad—with Jeremy, I mean. But that's not what he signed up for, is it? My dad was supposed to do that. But he didn't. But maybe Jeremy *can* end up being a caring father, or at least another friend. It doesn't sound like he is with his own kids, but who knows?"

We reached Jeremy's suburb after forty-five minutes. It was one of those vaguely familiar places whose name conjures alarming images from the back end of the national newscast—where periodically there is an especially senseless and spectacular murder. It was a sprawling, indefinite suburb. One strip of malls melded into the next. The town itself seemed to consist of fast-food outlets, motels, and a decent university. We followed Jeremy's directions to a quiet street of grim little ranch houses. We parked in front of the grimmest and littlest of all. That was Jeremy's. It was white, now smudged to gray. A chain-link fence surrounded the yard. A ruined car lay in the driveway. We got out of the car. It was at least ninety-five degrees, and the asphalt shimmered. It occured to me that this was not a place Jonas Salk would ever have lived. Jonas Salk would have been afraid to even drive through here.

Jeremy had warned me to knock on the left-hand door of the house and to be on guard for the pit bull that his neighbors kept in the right-hand side of the house. We walked up the driveway in silence. We could hear the pit bull barking insanely inside the house. Tom held Darian so that he was shielding him from the house: if the dog raced out, it would have to go through Tom to get to the baby. The house was a wreck. Windows were boarded up with plywood;

gutters drooped; siding was dangling off. I saw no left-hand door. Tom knocked hesitantly on the one door we did see, which was scraped and scarred. The dog's barking became more crazed. After fifteen seconds, there was the sound of dead bolts turning and chains being lifted, and the door opened on the scariest person I had ever seen. He was a white guy, maybe thirty years old. He was shirtless and heavily muscled. His head was shaved, and his mouth was open in a half smile, half grimace. This revealed that his top teeth were gold—all of them. Tattoos stretched across his chest and abdomen, covered his arms, shoulders, and neck. I know nothing about prison, but they looked like prison tats.

"Excuse me," Tom said. "We're looking for Jeremy?"

The guy stared blankly at us. "Jeremy?" he said. Even that one word was hard to understand, thanks to his impenetrable drawl and gold teeth. Three other guys, also all shirtless, muscled, tattooed, and frightening as hell, emerged from the gloom behind him and glared at us. Their eyes were freaky: they were amped up on something—crystal meth was my guess, given our location and their paranoia. The dog, if this was possible, was now barking even louder. It felt like a nightmare, *Deliverance* meets *Cujo*. The four of them looked us up and down. Tom and I had exactly the same thought: they were drug dealers, and they were trying to decide if we were cops. Thank God Tom was holding Darian, I thought. No cop carries a baby on the job.

Tom again asked for Jeremy. Finally one of them said, "Jeremy. You mean Jeremy Sampson?" Tom nodded, looking relieved. The guys relaxed a little, too. "He's around the other side of the house." The house looked too small to have an "other" side, but indeed it did. When we walked around to the left-hand side, past a broken umbrella and various crippled toys, we saw there was indeed a side entrance.

Tom knocked on the side door. It opened, and Jeremy stepped out to greet us. He said, "You must be Tom." Jeremy reached out, and the father and son embraced awkwardly, all shoulders and arms. So this was what happened when father and son met: nothing much.

Jeremy was in his late forties, but he looked fifteen years younger. He was wearing dress pants, dress shoes, and a garish blue Hawaiian shirt, untucked and open almost to the navel. Tom and Lana had the same first impression: he was a dead ringer for the "hero" in *Grand Theft Auto: Vice City*, one of Tom's favorite video games. (The aim of *Vice City* was to do anything to anyone for the hell of it—kill strippers, run over old ladies with your car, whatever got you off. Tom tried to block this image from of his mind: it was too alarming, maybe because it was too apt. What had Jeremy done with his life, if not whatever the hell he wanted?)

When I saw Jeremy, I finally understood why he had persuaded so many women to have his children. He had a shambling, raffish Dennis Quaid thing going. His hair was thick and dark, lustrous with some kind of product; he was always running his fingers through it, calling attention to its beauty. He had boyish, laughing features. His eyes were particularly striking: they were slit-narrow, but deep blue and twinkly.

Jeremy hugged Lana after he released Tom. Then he wiggled his fingers in front of Darian, who was delighted. He shook my hand and thanked me for coming. His voice was muddy like Tom's, but it also had an almost foreign lilt to it. He ended most sentences with a singsong "you know?" It made him sound a bit like a Mexican gangster.

He ushered us into the house and said, "Please, make yourself at home."

I didn't see how this was possible. For starters, the house was a sweatbox: it was ninety-five degrees outside and at least ten degrees hotter inside. There was no air-conditioning. A ceiling fan wheezed slowly, stirring the air hardly at all. We entered into what seemed at first glance to be a living room but was actually a bedroom, living room, dining room, and study all rolled into one. It took a while to figure this out, because the room was so gloomy: all the bulbs in all the lights were out, save a single, naked sixty-watt bulb on the ceiling fan. A blue carpet covered the floor. The shag was matted and covered with crumbs. Crushed Pokémon boxes, crumpled family photographs, dirty clothes, old copies of *National Geographic*, and books about American Indians littered the room. Crayon was scrawled on the walls. A boom box in the corner blasted Cambodian pop music.

Jeremy gestured for all of us to sit down. Lana perched precariously on a canvas director's chair that was missing an arm. I took a wobbly wood chair. Tom sat down gingerly on the trundle bed, which was covered in stained *Star Wars* sheets. Lana and Tom looked stunned. Lana was confused: she had always thought that doctors were rich. Tom was thinking, *This is my supersperm donor dad? I live ten times better than this.* I wondered how Jeremy expected us to spend the night here. The apartment already slept Jeremy, his old lady, and their two girls. Where would we fit?

As soon as we all sat, Jeremy popped out of his chair and stepped into the kitchen: "Would you like some melon?" he asked.

All of us nodded yes, for lack of anything else to say. Jeremy cut slices off an extremely ripe cantaloupe and handed them around. He wolfed his down, spilling juice all over the carpet. I watched as a cockroach strolled brazenly over to the juice spot. Jeremy tossed the rind on the dresser and forgot about it. Tom was sitting dumbstruck, Darian in one hand, melon dripping down the other. He was thinking, *This is not real. This is the like the dream I have where I win the lottery and they hand me a tangerine. First there were the drug dealers, now there is the melon, and it's ninety-five degrees, and I am here, and this is my dad.*

Jeremy stood up again and reached for the baby. "Do you want to give Darian a cold bath?" Lana smiled and said, "No, thank you." Jeremy asked again, "Don't you think Darian should have a bath?"

Again Lana said no. He asked again. Then again a few minutes later. Darian was squirming and fretful; Tom and Lana looked around dubiously for a place to put him down. Jeremy noticed and said, "I don't think he should crawl around here because of the cockroach problem, you know?"

We tried to settle in. Jeremy and Tom checked each other out surreptitiously. They shared a powerful brow, a big chin, and thick hair but little else. Even up close, Jeremy revealed surprisingly little of his age—some gray hairs, a chin that was beginning to wattle. His boyishness was astonishing. Tom was an eighteen-year-old with the air of a forty-eight-year-old. Jeremy was a forty-eight-year-old with the air of an eighteen-year-old. After so many wives and children, Jeremy ought to have looked dragged down by his troubles, but he didn't. He was careless, in both senses of that word. He was careless in that he didn't pay attention to the consequences of what he did—hence children and melon rinds strewn hither and yon—and he was careless in that he did not seem troubled by life's burdens. They didn't touch him. They were someone else's problem. He didn't seem malevolent, only puerile.

Carelessness made Jeremy a surprisingly gracious host. I would have thought he would be embarrassed by his house or weirded out by meeting an unknown son. But he seemed unperturbed. He joked, he punned, he flitted his attention from Darian to Tom to Lana to me. Tom and Lana seemed too overwhelmed to speak more than monosyllables. So Jeremy carried on a cheery, funny patter. Jeremy said a few words to Lana in Russian, then laughed about how he had once lived in Moscow for a few months and learned only how to curse. He was studying Japanese now, he said, and showed us his language tapes. He offered to buy Tom Russian tapes so he could eavesdrop on Lana's parents. He questioned Tom about what his mom was like. He said he was part Cherokee. He talked about the weather. Whenever there was a silence, he filled it, giggling his "you know?" at the end of every sentence. He was a natural-born seducer, and he was seducing us.

Jeremy delighted in Darian, and vice versa. He dandled Darian on his lap. He thrust stuffed animals into his face. He fanned him with a *National Geographic*. He handed Darian one of his books with an Indian chief on the cover. Darian grabbed the book and tried to eat it. "You wouldn't be so happy if he tried to scalp you, Darian! But you like to read. That's a good sign. Will you go to college one day, Darian? What do you want to do with your life, Darian?"

At the next conversational pause, Jeremy announced to Lana and Tom, "If you want to get married, I'll pay. We'll go down to the courthouse right now."

Tom broke into a big grin, his first relaxed moment of the visit. At last he had something to say. "We *are* married. We got married a few weeks ago." Jeremy grabbed Tom's hand with a huge pumping shake of congratulation. Jeremy offered to help Lana get her green card. He said he had a lawyer friend, they could fill out the paperwork that afternoon.

The silence descended once more. Jeremy asked again, "Do you want a cold bath, Darian?" Lana again said no, but this time Jeremy raced over to the bathroom and returned with a cold towel that he wiped all over Darian's face. The baby cried at the intrusion. "He's a crybaby," said Tom.

"Well," countered Jeremy, "he's just very expressive."

"I can't wait for him to talk," said Tom.

"That's the way parents are," Jeremy answered. "When the kids are young, they want them to walk and talk. Then, when they get older. It's 'Sit down and shaddup!' " He delivered the punch line as if he were performing in a nightclub.

Occasionally suspicion crept into Jeremy's conversation. He admired Darian's cuteness, then muttered, "The cuter they are, the more likely someone is to want to steal them." There was also an undertone of sleaziness. He advised Tom not to have two girlfriends at the same time, peculiar counsel to someone who was (a) your new son and (b) just married: "You get confused and call one of them the wrong name, and they both kick you out." He sounded as if he were speaking from experience.

After half an hour in the hot house, we were all sweating through our clothes, except Jeremy, who still looked crisp. Tom, Lana, and I were dreading the prospect of spending the rest of the day there. We had to escape. I suggested we get some lunch.

Over Cuban fast food and in the air-conditioning, everyone relaxed. Tom asked Jeremy if he had ever expected to meet his sperm bank kids. In a loud voice, Jeremy started to answer, "I never knew there were any sperm bank kids." Midway through the sentence, he remembered he was in a public place, looked around campily, and dropped his voice to a whisper. Tom and Lana laughed. Jeremy buzzed Tom with questions, sometimes interrupting answers to make a joke or do an impersonation. What's your favorite video game, Tom? Why do you like it? Do you play chess? Checkers? What's your favorite drink? Jack and Coke? Really? What happened when you found out you were a Nobel sperm bank kid? Did you ever break any bones? Which ones?

Tom enjoyed the attention, but I was uncomfortable. Jeremy seemed superficial. Not fake, exactly, but theatrical. He listened to Tom's answers, only enough to ask the next question. He didn't seem to care what Tom was saying. It felt like a show of affection for Tom's benefit. But if it was, so what? Was faked affection worse than none?

Jeremy picked up Darian and stared at his chunky cheeks. "Look, he's Marlon Brando!" Tom loosened up, too. He called Jeremy "Grandpa." Jeremy smiled at this in a peevish way. Jeremy scooped up Darian and paraded him around the restaurant. The cashiers cooed over the baby, as Jeremy beamed. Tom whispered to me, "It's what I was hoping for. It's good. I feel comfortable."

It was obvious that Jeremy was no genius. But it was also obvious how he had persuaded Julianna McKillop and Robert Graham that he deserved to be a donor to the Repository: he had a gift for making people feel at ease, and he had a quick tongue. He could charm the pants off anyone (and often had). Twenty years ago, when he was a medical student with a pretty wife, before his life got so messy, he must have shone with all the promise in the world.

I announced that I was staying in a hotel and offered to get a room for Tom and Lana, too. Jeremy looked relieved. We found a Marriott with a pool. As we checked in, Jeremy pulled out his wallet and tried to hand me a hundred-dollar bill to cover Tom and Lana's room. I refused it, so he thrust it at Tom. "C'mon, Tom. You had to pay for the airline ticket, right? David won't take it, so I have to give it to someone, you know." Tom reluctantly accepted the C-note. As he took it, Jeremy said softly to him, "Remember this when I am old and broke and retired." In case Tom hadn't heard, Jeremy immediately said it again, as a question: "You'll remember this when I am old and broke and retired, right?" Later, when he knew Tom was watching, Jeremy picked up Darian and said, "At least you'll take care of me when I'm old, right, Darian?" It didn't sound as if he was joking. Tom was embarrassed. There was something sad about a man with so many children hoping a $100 gift would persuade his sperm bank son to cover his nursing home bills.

At 4 P.M. we had to pick up Jeremy's two kids—or rather, the only two of Jeremy's many kids who lived with him. I drove Jeremy to the babysitter's. He thanked me for suggesting the hotel room. "You don't want to sleep on that floor, not with our cockroach problem." I asked him about the scary guys next door. They were his landlords, he said. "The guys, they don't really seem to do anything." He said this in a way that made it clear that they did something but he was afraid to say what. In front of Tom, Jeremy hadn't wanted to talk about why he was living in such squalor, but he opened up a little bit when we were alone. His job paid okay, he said, but he was a civil servant, not a rich doctor in private practice. He had to give half his modest income in child support—half was the maximum allowed by law—for his various kids. "Yeah, it's not really the best living situation, you know. I don't have much left over after all the child support. That's what you get for having X kids, I guess." But he didn't sound too regretful when he said this—that carelessness again.

Jeremy's two girls were playing in the yard when we arrived. Mimi was nine; Stacy was eight. They were beautiful and brown-skinned—their mom was Haitian—with their dad's thick hair and bright eyes. They were darlings: Stacy was powerfully built and full of energy. Her older sister was lither and a little calmer. They bounced all over the car, played with Jeremy's hair, teased each other and their dad.

When we arrived back at the hotel, the girls were excited to meet Tom, Lana, and Darian but more excited to swim in the pool. They had only the fuzziest idea of who Tom was and why he was there. At first they thought Tom was their uncle. Jeremy finally managed to explain that Tom was their brother, which did not surprise them; they had so many brothers already.

We spread out around the small hotel pool. Lana lounged in a beach chair with Darian. The girls took a shine to Tom, though I suspected they would take a shine to anyone who paid attention to them. Tom loves kids, and he found it easier to talk to them than to Jeremy. He raced them across the pool and played Marco Polo. Jeremy joined them for a game of keep-away. When everyone was exhausted, we sat around the table and Tom gave the girls arithmetic problems while they played peekaboo with Darian. "Can the baby stay with us?" Stacy asked. Everyone was laughing and goofing. It was early evening by now; the sky was pink and hazy and soft. The vicious heat had dissipated into an easy warmth. Tom was calmer than I had ever seen him.

Jeremy, wearing a straw hat and an unbuttoned white cotton shirt, bounced Darian on his knee and gazed at his kids—Tom and the girls—with a bemused smile.

Tom said he was interested in being a writer, too. He asked Jeremy about the book he had written. Jeremy looked embarrassed. "It was self-published," he said. "It was about a dream I had." He didn't elaborate. Then he said, "I was thinking of being a writer, but it's really hard to get published, really hard to break in. So I gave up on that." I asked him about his current job. He said he liked it because it was easy working for the state: "When I graduated medical school, I worked for a private practice, and I was going to set up my own practice. But I looked into it, and it was just so complicated. You had to handle billing and a secretary, and you had to find patients and advertise. It just seemed like it was going to be too much."

He knew my magazine *Slate* was owned by Microsoft, and he wondered if I had any Microsoft stock options. "I don't own any stocks," Jeremy said. "I think I could be a really good investor, but I don't know enough to do it right. What I have always planned to do is make a practice portfolio and then try it for a while and see how I do, and then when I learned how to do it right, I would invest real money. But I haven't done that yet." My God, I thought, he sounded exactly like Tom did in the morning, when he was considering and rejecting possible careers. Was this Jeremy's genetic gift to Tom— this indecisiveness, this giving-up-before-you-startness?

It was time for dinner. We couldn't all fit into my rental car. Tom volunteered to ride with Jeremy. (Jeremy's car looked like his house: trash festooned the floor, books about American Indians were piled on the seats.) We stopped to collect Jeremy's purse-lipped old lady (though wouldn't you be purse-lipped if you lived as she had to?).

We ate at a small Thai restaurant. Tom and Jeremy were next to each other, and I sat across from them. Their gestures were eerily similar. Each leaned slightly back in his chair, right elbow on the table, left hand crossed and resting in the right elbow's crook. When they talked, they tilted their heads a few degrees to the right.

"Are you happy?" Tom asked his dad.

"Yeah, I guess. I am too busy not to be happy," Jeremy answered. Then he turned the question back at Tom. "You look happy, Tom. You have a beautiful kid, a good job, a pretty wife. You should be happy!"

Tom smiled. Jeremy described a few of his other kids, particularly a son who had gotten straight As and gone to math camp.

"I did real bad in high school," said Tom. "All Bs. I didn't study. I was really lazy."

"You sound like me," said Jeremy.

"I really don't know how to work hard yet. I thought I would learn how to work hard in college, but I haven't really," said Tom.

"Well, there are worse things than being lazy," Jeremy said.

Jeremy suggested we get doughnuts for dessert, so we convoyed to a nearby Krispy Kreme. Tom and the girls were mesmerized by the doughnut machine and hyper at the prospect of eating all that sugar. Tom was in an expansive, generous, gleeful mood. He insisted on buying everyone doughnuts, dozens of them, all kinds of them, far more than we could eat in days. We crammed into a couple of small tables, cutely pressed up against one another. Tom fed doughnuts to the girls. Jeremy nibbled slowly on a small glazed; the girls leapt up and down in a sugar mania.

Jeremy suddenly asked me, "Where would you want to live if you had a hundred million dollars?"

"I don't know," I answered. "Maybe Los Angeles in winter, Vermont in summer, New York City the rest of the year."

"I would live in the Queen Charlotte Islands, way up off the coast of Canada. There's hardly anyone there, just some Indians."

"Why are you so interested in Indians?" I asked.

"I think it's the yearning for a simpler life." He looked at the girls and at Tom. "But that will never happen."

We returned to the hotel. Jeremy and his family said good night and left us. Tom changed Darian into his pajamas and reviewed the day—the drug dealers, the melon, the heat, the filth, the pool, the doughnuts. Mostly, Tom said, he was really happy: he felt pretty comfortable with Jeremy. He was glad that Jeremy was curious about him. He didn't care that his genius sperm bank father wasn't Jonas Salk.

Still, something was eating at Tom: Why had Jeremy fathered so many kids, and why had he left them?

"I can see why he got so many girls to go to bed with him. He has all this charisma. But I was trying to figure out the kids, and it just hurts my head. He really looks like he loves Mimi and Stacy. If he felt that way about the others, I don't know how he ever left them. Is it that he just has bad taste in women? I can't believe that. I can't believe he has X kids and married that many times. The type of person who would do that, it just doesn't seem like him."

"People make mistakes," Lana offered.

"Yeah, but not X times," Tom cracked.

"Or maybe he just has bad common sense, like you," Lana said.

"Yeah, maybe. It's kind of scary that I am like him, because he has done so much stuff I don't want to do—the kids, the marriages."

Jeremy and his daughters returned to the hotel in the morning. We spent a few more hours poolside. Jeremy was wearing his battered straw hat, a white linen shirt, and a towel wrapped around his waist like a skirt. He looked like a contented old hippie. Jeremy and Stacy chicken-fought in the pool with Tom and Mimi. The girls pushed Darian around in his stroller when he fussed. "You are wonderful aunts," Tom called to them. When Tom tried to feed Darian a bottle, Stacy grabbed it from Tom and rebuked him: "Don't play with my baby." Jeremy danced around the edge of the pool with Darian. He stopped for a moment and asked, "How many kids do you want, Tom?"

"Two of them is all I want," said Tom, "and we're going to wait five years for the next one."

"Two or three, that's a good number," said Jeremy. "More than that, and the possibility of fighting increases exponentially. There are too many different combinations."

"Jeremy, can you write down a list of *all* my brothers and sisters and their ages for me?" Tom asked.

"Yeah, I can do that. I used to have a list like that around. But they keep on coming out of the woodwork. There are more every day. You and Alton and—"

I asked Jeremy the question Tom had asked me the night before: Why had he had so many kids? If he could live his life over, would he have them again? He thought for a moment, then answered, "I don't know. Stacy and Mimi are number X minus 2 and X minus 1. So imagine, if I had stopped earlier, then I would never have had them."

But what about all those other kids, all the ones you don't see and don't take care of? What about them? That's what I wanted to ask, but I didn't.

Jeremy proposed we drive to South Beach. It was a beautiful day, Tom and Lana had never seen the ocean, and that way Jeremy could drop us off at the Miami airport late that night. Jeremy and Tom drove together. On the way down, with me out of earshot, Jeremy congratulated Tom on finding a foreign wife. Foreign girls, he said, let you get away with a lot more. You can mess around with other women and then explain to your wife that cheating is the American way. He told Tom that he even had a built-in excuse if he got caught. "You can tell Lana, 'Oh, it's normal for me to want two girls. I was raised by my mom and my sister, so I need to have two women in my life.' " Tom was revolted. He had been married for two weeks, and his dad was offering him tips on how to cheat. Tom changed the subject.

Tom and Lana were agog at South Beach—models in bikinis, fortune-tellers, stores selling water pipes and incense, outdoor bars, street musicians. Jeremy found a shady spot on the sand, beneath a lifeguard shack. We laid out towels and settled in for the afternoon. It was mellow and sweet and happy. Stacy fed Darian a bottle. Mimi braided Lana's hair. Jeremy changed Darian's diaper. Jeremy and Tom waded in the surf together. Jeremy came back first. We were alone. I asked him what he thought of meeting his Repository son.

"It's going better than I expected. He's quite a bit like me. He's kind of an intellectual, likes college, likes learning. He looks like me. We have the same way of talking, sort of. He is easy to talk to, not too quiet, not too talkative. I hope we will see each other again. I don't see why not. Maybe I'll move to the Midwest."

Jeremy asked whether, if they made a movie from my book, he could sell his movie rights and get some cash. I told him I doubted that anyone would make a movie, so he shouldn't count on anything. "Yeah, it needs conflict," he said, then added jokingly, "Where's the conflict? I guess I could kidnap Darian for two weeks or something and ask for a million-dollar ransom."

The kids returned, and we ran races around the lifeguard shack. Tom refereed wrestling matches between his two "sisters"—as he was now calling Mimi and Stacy. He buried them in the sand, then dug them up. He grabbed them by the wrists and spun them around like helicopter blades. Jeremy suggested we go visit another one of his children, who lived not far away. Tom nixed the idea: this was family enough for him today. Jeremy marched around with Darian, singing "Supercalifragilisticexpialidocious."

It was dinnertime, and we had to start heading toward the airport. Tom stopped at a gift shop to buy presents for all of us: candles for Stacy and Mimi, a Rastafarian hat for me, and—because he had seen how much Jeremy liked Indians—a picture of a tribal chief in a feathered headdress for Jeremy. We found a pizza parlor. Darian sat on Jeremy's lap and banged the table. "He's a drummer, I knew it!" Jeremy exclaimed. Jeremy showed off the baby to the cooks and other diners. He was calling himself "Grandpa" without any hesitation.

Jeremy and the girls drove with us to the airport and decided to wait until we boarded our flights. On the shuttle from the rental car lot, Tom whispered to me, "I am really, really, really happy." He and Lana held hands and nuzzled. Tom and Jeremy made plans for Jeremy and the girls to come visit Kansas City soon, probably before Thanksgiving. We sat in a circle on the floor of the airport terminal and played cards. Tom called his mom. They talked for a few minutes. Then Tom handed the phone to Jeremy. The first thing Jeremy said was, "Hi, this is Tom's dad." Then he added, "Thanks for letting Tom come out to see us." Jeremy listened a little and gave the phone back. Tom then passed the phone off to Stacy, who said a few words. (Later, Mary recounted her end of the conversation. She was annoyed: Mary had told Tom, "Here, talk to your sister," and passed her phone to Jessica. Tom had heard "sister," and given his phone to Stacy. Mary said, "Jessica was so hurt when I gave the phone to her and it turned out that he had done that, since Jessica is his sister, not the strangers he is so quick to refer to as 'sisters.' ")

My flight left first, so I hugged everyone good-bye. As I waited in the security line, I gazed back at them. They were playing cards. What were they, I thought, if not a family, even an all-American family? There was Tom, who was too adult for his eighteen years; Tom's childlike father, Jeremy, the careless "genius" sperm donor; Tom's immigrant wife, Lana; Tom's half sisters Stacy and Mimi, merrily oblivious to the paternal abandonment that might await them; and, asleep in his car seat, Tom's own son, Darian, the heir to Nobel sperm bank genes, which is to say, the heir to God knows what.

CHAPTER 13

SPERMICIDE

The Nobel sperm bank celebrity grows up: Doron Blake, age seventeen. *Dave Luchansky/ Getty Images*

One question still puzzled me: Why had the Repository for Germinal Choice closed in 1999? I couldn't figure it out. Robert Graham had considered the sperm bank the most important work of his life, and he had intended it to survive him. Yet two years after his death in 1997, the bank was finished. Surely Graham hadn't simply abandoned it. What had happened?

I finally learned the whole story from Anita Neff, the Repository's final manager. Anita had been a shadowy presence throughout my research. Lots of the donors I met had been recruited by her. She had played the heavy in the Donor White saga, cutting the communication between Beth and Roger in 1997. But no one knew what had become of her. One donor thought she had moved to Italy. Another said she was in Norway. For two years, I hunted her unsuccessfully. Then, in 2003, the documentary researcher Derek Anderson found Anita in southern California. After several years overseas, she had settled just a few miles from the Repository's old office in Escondido. After some cordial e-mails, Anita agreed to meet me on my next trip out west.

We made a date for brunch at a Del Mar mall. There was no one who looked remotely like a former sperm bank manager at our meeting point. Finally a woman approached me and said, "I'm Anita. Are you David?" Anita was stunning. She was wearing a black skirt, slit to reveal all but a few northern inches of her excellent legs, black lace stockings, a tight black top with a fur collar and fur wristbands, and an elegant, wide-brimmed black hat. She was forty-five years old, but she looked thirty. (When I saw her, I understood why donors had spoken of her with such, um, reverence. One had remembered with evident delight how "she was motivation when I was doing my donations, if you know what I mean." Not surprisingly, she had been a superb recruiter.)

Anita had not spoken publicly about the Repository since she had left it, but she was glad to have the chance to talk. She was direct and funny. One reason I particularly liked her was that she was not a eugenics zealot, unlike others who had worked for Graham. Quite the contrary. Anita was the ninth of eleven children, and she didn't have any kids herself; she did not spend her days fretting about the shortage of children in the world. She said she had ended up at the Repository by chance. In 1993, when she was an HIV and pregnancy counselor, she had answered a classified ad for the Repository.

She had never heard of the sperm bank, but Graham had charmed her during the interview. Anita didn't care about Graham's genetic crusade, but she didn't see anything wrong with the bank's mission, either. And she was really excited about the challenge of repairing the bank. When she arrived in 1993, the vaunted Nobel Prize sperm bank was a mess. Under a series of managers in the 1980s and early 1990s—none of them technicians or scientists—the bank had slid into a sorry decline. It had managed to spawn a lot of kids, but its methods were haphazard, to put it mildly. Graham, for example, had shamefully loosened his standards for donors. Graham had begun with the intention of recruiting only the best men in the world. But by the mid-1980s, despite occasional catches such as Donor White and Donor Fuchsia, he and his managers were accepting almost any man who walked in the door. A series of volunteer donors, moderately accomplished egotists such as Jeremy, had snowed Graham. They had sold him on their brilliance, sometimes lying to him about their IQ, athletic achievements, musical talents, and professional achievements (lies that were then repeated in the donor catalogs). Forget about Nobel laureates; the Nobel sperm bank was taking men you wouldn't wish on your ex-girlfriend. (Graham had more quixotic ambitions, too: he tried to enlist Prince Philip of England as a donor, despite the absence of evidence that the prince has ever engaged in any kind of cognitive activity. And according to Sylvia Nasr's A Beautiful Mind, schizophrenic mathematician John Forbes Nash also considered donating his genius sperm, though I found no evidence that Graham had contacted him.)

It seemed to me that the Nobel sperm bank ultimately became a kind of scam. Its reputation for Nobelists and geniuses fooled mothers into thinking they were getting a better product than they were. They expected Nobelists, and ended up with men like Jeremy. But the decline in quality was invisible to customers. The catalogs remained grandiose, duping women into thinking that the donors of 1988 were the same kind of men as the Nobel donors of 1980.

The Repository had other dubious practices. Most sperm banks limited themselves to donors under forty, because older men's sperm suffer from more genetic abnormalities. Not the Repository. Donor White, for example, was over fifty when he was recruited and nearly sixty when he stopped giving. Graham also played loose with child limits. He originally restricted donors to twenty offspring, to reduce the possibility of accidental incest. But when Graham found himself short of semen—a regular problem—he relied repeatedly on his most prolific donors. One donor fathered twenty-five kids, according to Anita. I suspect gold medalist Donor Fuchsia fathered even more than that. (Why? Because of the thirty kids I know, eight are from Fuchsia. That's 27 percent of them. My sample could be statistically anomalous. If it's not, that suggests that more than fifty of the bank's two-hundred-odd kids came from Donor Fuchsia.)

Anita said she had whipped the Repository into shape. She had insisted that it follow the American Association of Tissue Banks' guidelines for sperm banks. That meant declining donors who were over forty or had fathered too many children. She tightened the application process. She added new blood tests and disease screenings. She flew donors to San Diego for a brutal physical. Before Anita, few prospective donors had been rejected on health grounds. Under Anita, that happened all the time. Because donors did their business at home, she DNA-tested all sperm samples to make sure they were sending their own seed. Anita brought truth in advertising to the Repository. The catalogs were toned down; donors' accomplishments were not exaggerated. If a donor had scored only 420 on the verbal section of his GREs, his listing said so. Anita reorganized the records after discovering to her horror that the bank had reused color identification codes (two Donor Oranges!). Anita also cracked down on the ad hoc arrangements some donors had struck with previous managers. One donor, for example, intended to pay for his offspring's college educations. Anita said no. "That was making the bank play a role it was not supposed to play. It was causing people to choose their donor based on the hope of getting something out of him. That was not what Dr. Graham wanted."

Anita and Graham shared recruiting duties. Anita pursued different kinds of men than her boss did. Graham still hungered after great brains, even if he couldn't find them. Anita prioritized health over genius, making sure every man she found had a stellar medical history. And she broadened the donor pool. Graham let her "recruit highly qualified people that he would never have recruited." In other words, nonwhite people. She enlisted a Samoan and a Native American. (Anita thought Graham was not quite the racist he was made out to be. "He was willing to be convinced that the world was full of good people of high capacity, and they don't all have to be German. He was willing not to always hold on to the past.")

The Repository sustained its popularity during the early and mid-1990s. The waiting list reached eighteen months, because there were never enough donors. Usually, Anita could supply only fifteen women at a time with sperm. California Cryobank, by contrast, could supply hundreds of customers at once. Demand at the Repository remained strong even when Graham started charging for sperm. In the mid-1990s, the bank collected a $3,500 flat fee per client, a lot more than other banks. Ever the economic rationalist, Graham had concluded that customers would value his product more if they had to pay for it.

The press remained enthralled with the Nobel sperm bank, in a delighted and revolted way. When William Shockley died of prostate cancer in 1989—scorned, loathed, and bitter—his obituaries noted, contemptuously, that he had been the Nobel Prize sperm bank's flagship donor. In 1991, the Annals of Improbable Research awarded Graham its first Ig Nobel Prize in Biology. Even so, newspaper and TV reporters streamed to Escondido to interview Graham and Anita. They clamored to know whether the dream had come true. Were the kids special? Were they "wonder kids," "designer babies," "superchildren"?

Graham didn't know the answer. He had intended to use his Repository kids as lab rats in scientific studies of their abilities, but that plan, like all his others, had not turned out as he hoped. In 1992, he had mailed an initial survey to all the parents, asking for basic information: How old were their children? How well were they doing compared to other kids? Any IQ test results yet? Practically no one had returned the survey. The parents had rarely shared Graham's eugenics bent to start with, and they cared even less once their babies were born. Graham was disappointed that he would never prove that his sperm had produced better children.

Graham was eighty-six by the time Anita met him but still spry. A bout with cancer slowed him down a little and he wore hearing aids, but he was otherwise the same polite, energetic, flirty man he had been thirty years earlier. He would swim at his pool in the morning, then visit the Repository for an hour in the afternoon. He showed off to Anita—in a cute way, she thought. He was writing his memoir, and he wanted to include lots of pictures of himself as a young man, including several in scanty bathing suits, because "he was so proud of his physique." Graham loved being around pretty women, but he was gentlemanly, not lecherous. He liked to call Anita "my Varga Girl." (One donor recalled that when the eighty-eight-year-old Graham had come to recruit him, he had asked Graham if his wife, Marta, missed him while he was away. Graham had looked slyly at the donor, proudly hitched up his pants, and proclaimed, "Well, I took care of that before I came out here.")

In the 1990s, Graham accumulated lifetime achievement awards from ophthalmologic and optometric societies. He accepted the prizes graciously—he was incapable of doing anything ungraciously—but they meant nothing to him. He was still a crusader for his real cause, Anita said. "When he was probably ninety, he was getting a big award from an ophthalmology association, and I asked him how he felt about it. He said, 'I really don't care. That's the past. I care about the future, about what I am going to do in the next ten years.' "

Even in his late eighties, Graham was leafing through *Who's*

Who looking for donors, sponsoring a conference on human genetic engineering at UCLA, eagerly granting TV interviews about the Repository, happily trekking cross-country to flatter men one third his age into becoming donors.

Graham was desperate for the Repository to continue after his death, Anita said, but he had a problem: none of his children was interested in it. In fact, they hated it. Who could blame them? Graham sometimes seemed prouder of his sperm bank offspring than of some of his own children. Graham's wife had no particular fondness for the bank, either, Anita said. She tolerated it because she loved her husband.

Meanwhile, the Repository was bleeding cash. Graham had hoped that customer fees would enable the bank to break even. But the fees were never remotely enough. According to its tax returns, the Repository spent about $170,000 a year to collect and distribute sperm but garnered only $20,000 from customer fees, $40,000 in a good year. The difference had to come from somewhere: namely, Graham's pocket. He shelled out $100,000 or more every year to keep the bank, going. As long as he lived, that was fine—he had millions to spare. But what about when he died?

In 1994, Graham thought he had found the perfect solution: an Ohio magnate named Floyd Kimble. A World War II veteran, Kimble had settled in eastern Ohio after the war and thrived in all kinds of businesses. He drilled gas wells; mined limestone, coal, and clay; operated a landfill; manufactured cement mixers; farmed; and more. He succeeded at everything. His biggest score came in 1988, when he won a $600 million lawsuit over natural gas contracts. It's not clear how much of that he was actually paid, but the next year, he and his wife, Doris, set aside $30 million to start the Foundation for the Continuity of Man. The name was a conscious echo of Robert Graham's Foundation for the Advancement of Man.

Kimble was a Graham acolyte, but he was cruder in every way than Graham. Kimble's manner was rougher; his eugenic ideas were less sophisticated; his prejudice was more overt. Still, the two men had plenty in common: they were self-made millionaires from rural backgrounds, and they shared a can-do spirit and the conviction that our genes were going to hell.

Kimble had slightly different goals than Graham. His ideas were a queer stew of Graham's eugenic alarmism and his own apocalyptic environmentalism. As a farmer, Kimble believed that hybridization was weakening plants even though it increased crop yields. Kimble theorized that hybridization had weakened humanity in the same way: Too many unfit humans were being born, and they were breeding too much. Kimble also dreaded a global environmental catastrophe. His solution to all this: rather than *distributing* great sperm à la Graham, Kimble decided he would *preserve* it for the future in a remote, radiation-proof vault. Hence, the Foundation for the *Continuity* of Man. In 1991, Kimble bought a real bank building in suitably far-flung Spokane, Washington, and set about stocking the vault with human semen. He also planned a separate warehouse for plant seeds and animal semen.

Kimble contacted Graham to discuss the genetic crisis and the Repository, and they struck up a friendship. Through his connection with Graham, Kimble met the former Repository manager Dora Vaux and hired her to run his bank. Vaux solicited Repository donors to contribute to Kimble's storage bank as well. (Michael the Nobelist's son says he gave to both banks.) Kimble also invited more than fifty Norwegian athletes to donate, though apparently none accepted the offer. For Kimble, Norwegian athletes were as good as it got. In 1996, Vaux told a reporter that "racial purity" was a goal of Kimble's bank, that she was collecting only from "high-achieving white men." Vaux also said that if sperm from black men were ever collected, it would be stored separately from the white sperm.

Graham realized that Kimble could ensure the future of the Nobel sperm bank. He was the perfect heir to it. He was twenty years younger than Graham, he had plenty of money, and he believed in genius sperm banking. Kimble was certainly a better bet than Graham's own family, which would shut the bank the first chance they got. According to Anita, Graham and Kimble struck a handshake deal. Kimble replaced Graham as the cash cow and agreed to continue the Repository when Graham died. During the last week of 1994, Kimble donated $400,000 to Graham's foundation, enough to run the Repository for three years. When that money ran out, Kimble supplied another $100,000.

Despite Kimble's cash, Graham still got to supervise the Repository, and he had every intention of doing that for years. In June 1996, he celebrated his ninetieth birthday. In February 1997, Graham traveled by himself to Seattle for the annual meeting of the American Association for the Advancement of Science. "He was on a recruiting trip, of course," said Anita. "He would go to meetings like that to walk the halls," scouting for good-looking young hotshots. On February 13, Graham fell in the hotel bathtub, hit his head, and drowned. His *New York Times* obituary gave equal billing to his invention of shatterproof eyeglasses and his Nobel sperm bank. *Time* magazine marked his death with an item in its "Milestones" section. (A couple of items below, "Milestones" also noted that week's conviction of biologist Carleton Gajdusek for child molestation. Gajdusek was a Nobel Prize winner. I wonder if Graham had ever asked Gajdusek to donate his Nobel sperm. What would he have made of Gajdusek's crime? Would he still have coveted Gajdusek's Nobel seed?)

About three hundred people attended Graham's funeral at Emmanuel Faith Community Church in Escondido. Soon after, Anita mailed a letter to the donors announcing Graham's passing. "We wanted to send you this personal note since you are a very special part of his dream. As per Dr. Graham's wishes we will continue to operate the Repository in the same manner as in the past."

But of course it wasn't the same. Graham's will didn't mention the bank, but Floyd Kimble, as expected, assumed responsibility for it. He provided cash when Anita needed it. But after a slow year, Anita was ready to move on. She resigned and moved to Europe with her husband. Kimble hired no permanent manager to replace her. Then, in September 1998, Kimble died suddenly at age seventy.

Kimble's death sealed the fate of the Repository. He apparently had made no provision for the bank in his will. With Anita's departure, there was no one to collect and distribute sperm. And with Kimble's death, there was no one to pay for it. In early 1999, Robert Graham's widow, Marta, Floyd Kimble's son, Eric, and medical director Frank Andersen decided to shut the Repository for Germinal Choice. After nineteen years and 215 children—not one of them a Nobel baby—the Nobel Prize sperm bank would go out of business.

On April 29, 1999, Andersen announced the shutdown in a letter to donors. He told them that he was arranging for "proper clinical disposal" of the stored sperm. If donors wished to collect their samples for "personal use," he would try to arrange it, but "you should know before considering such a course that it may be difficult or impossible to find a facility to accept and store the specimens, and that the cost of such an effort would be considerable."

Donor White tried to save the bank, in his own unobtrusive way. He thought if the shutdown were publicized, some rich man might step in to save it. Roger leaked Andersen's letter to a San Diego TV station, but it refused to cover the story unless he appeared on camera. Careful of his privacy, he wouldn't. Instead Roger e-mailed Logan Jenkins, a *San Diego Union-Tribune* columnist, telling him, "I have the feeling that if Dr. Graham were still alive, he would not wish to see his work ended in the manner proposed." Jenkins did write a column. It ever so slightly regretted the bank's demise, but Jenkins also did what journalists had always done to the bank— mocked it: "The world's most notorious sperm bank is undergoing a radical vasectomy. The Repository for Germinal Choice, the mercilessly teased brainchild of Dr. Robert Graham, is tying its tubes after helping produce a litter of 215 genetically boosted babies." So the

Nobel Prize sperm bank died as it had lived, half science, half comedy. Practically every newspaper and TV station in the United States had covered the bank when it opened. Jenkins's column was the only thing written about it when it closed.

No sperm sugar daddy stepped forward to rescue the bank. The shutdown proceeded as planned. None of the donors requested his sperm back. Marta Graham asked Steve Broder, Graham's original technician, if he wanted to buy any of the Repository's equipment for use at California Cryobank. Broder didn't, but he volunteered to help supervise the closing. One morning in mid-June, Broder and Marta Graham descended on the Escondido office for the final time. The Repository's records—just a bunch of file folders—had already been entrusted to the Repository's office manager. (I don't know where she keeps them; she never answered my queries.) A medical waste company arrived. The liquid-nitrogen vats, the background of hundreds of pictures of Robert Graham clouded in vapor, were emptied and carted away.

Next, spermicide. The frozen vials—once so precious that they had been double-locked and shielded by lead, that reporters had begged for a glance at them, that women had traveled around the globe to get their hands on them—were dumped unceremoniously in red biowaste bags and driven off to the incinerator. Dr. Graham's dream began in ice and ended in fire.

Neff wasn't nostalgic when she recounted the end of the bank. "Sperm banking will be a blip in history," she said. The Nobel sperm bank, she implied, would be a blip on that blip. And in some ways, she is clearly right. The Repository for Germinal Choice pioneered sperm banking but ended up in a fertility cul-de-sac. Other sperm banks took Graham's best ideas—donor choice, donor testing, and high-achieving donors—and did them better. They offered more choice, more testing, more men. And they managed to do so without Graham's peculiar eugenics theories, implicit racism, and distaste for single women and lesbians. The Repository died because no one needed it anymore.

But the dream the Repository represented is more alive than ever. Since my two children were born, I have been thrust into the world of yuppie parental ambition. Child making and child rearing have become full-contact sports. Parents start enriching their children in the womb and never stop. The amount of parental involvement in children's lives is scary. We dose them with Ritalin and antidepressants in the cradle, use Machiavellian maneuvers to enroll them in honors classes and select soccer teams. We live by a competitive creed: We must give our children any edge we can.

For the moment, we seek advantage through drugs and classes and tutors, but we will use genes as soon as we can. The Repository's notion—that good sperm will make good children—is too crude for our age,* but more sophisticated science is coming, advancing Graham's dream to the twenty-first century. The first hints of the new world are already here. A technique called "preimplantation genetic diagnosis" (PGD) allows a doctor to run genetic tests on eight-celled embryos created by IVF. The doctor and parents can then select the most genetically fit embryo for implantation in the womb. At the moment, PGD can screen for only a few genetic diseases (as well as for gender), so it's used chiefly to help parents protect their kids from dread ailments such as cystic fibrosis. But eventually PGD will be able to tag genes associated with musical ability, blue eyes, or intelligence. When that happens, most parents will still reproduce the old-fashioned way. But the few who really care about beating Mother Nature—the ones who wrote to Dr. Graham in 1980 and who shop for egg donors at Harvard today—will be lining up for

---

* Well, maybe not too crude for everyone. A British production company recently announced plans for a reality TV show tentatively titled *Make Me a Mum*, in which one thousand men will compete to become the sperm donor to a first-time mother by showing off their intelligence, good looks, and health.

PGD and hoping for a prodigy. The old-time eugenics of Graham and Shockley and Galton is dead. No one cares about the national "germ plasm" anymore. But private eugenics has arrived to replace it. If we can get better genes for our own kids, many of us will do so. Just like the first Nobel sperm bank customers, we are captive to the great delusion that we can control our children, that we can make them what we want them to be, rather than what they are.

The question I am most often asked is: Did the Nobel sperm bank work? By which questioners mean: Did it make superkids?

I don't have a simple answer. Of the 215 children of the Nobel sperm bank, I know of 30, aged six to twenty-two. I've met some of them, talked to many of them, and e-mailed with others. In some cases, I have only talked to their parents about them. My sample is not random; these are families that contacted me. They are probably exceptional in all kinds of ways. Most of them, for example, are single-mother families. Intact families tend to be less open about their DI secrets, partly to guard the relationship between father and children. I also suspect that my sample families are *more* satisfied with the Repository, because people are usually more willing to talk to reporters about things they are happy about. Still, let me try to sum the children up.

A few of them—Alton Grant, for example—have brilliant minds. A few others have wonderful physical talents: there are a couple of superb dancers and at least one amazing singer. Of the rest, most are very good if not great students. Several kids perform below average in school. Almost all are in excellent health, but one boy in the group is autistic and one girl suffers from a debilitating muscle disease. In short, they are certainly above average as a group, but the range is very wide.

Is this a tribute to Robert Graham and his great sperm? I don't know, but I doubt it. These are fortunate children: they come from prosperous homes—middle class and up—and they have exceptionally attentive mothers. Most children would thrive in such surroundings. Measuring what the sperm donor contributed is simply impossible. Yes, the smartest of the kids had smart donors, but they also have smart mothers, and they have been raised in intellectually challenging environments. The most physically gifted had physically gifted donors, but they also have physically gifted mothers, and their parents have cultivated their talents. So the question of what the Repository gave its children is unanswerable. Though I suppose it could be answered this way: of all the parents I talked to, only one regretted using the Repository. The Nobel sperm bank may not have met the world's expectations, but it met the expectations of those who mattered most: its customers.

The children of the Repository for Germinal Choice have certainly *not* become an elite, celebrated cadre, as Graham hoped in his most ambitious moments. All the children lead very private lives—except one: the prodigious Doron Blake, the Repository's second child and its most celebrated public legacy. Doron's early achievements—computers at two, *Hamlet* at five, IQ of 180—and his mother's publicity seeking made him the bank's adorable mascot. In 1995, when Doron had reached the venerable age of thirteen, Graham declared, "When Doron Blake is old enough, I'm going to ask him to become a sperm donor himself at the Repository." But within two years, Graham was dead. By the time Doron turned eighteen in 2000, the bank was gone.

The Repository died, but the fascination with Doron lived on. Reporters kept calling him to find out how Graham's experiment had turned out. As a prodigy and as practically the only Repository child who talked openly to the press, Doron was a precious commodity. He turned the media interest into a nice income stream. Any reporter who wanted to talk to him had to pay. Before he turned eighteen, his mom, Afton, dunned reporters and deposited the proceeds in a college fund. Now that he is an adult, Doron has taken over the business, asking $500 and more per interview. He uses the cash to clear college loans, buy books, pay for vacation travel. Doron told me in 2001 that he had performed his sperm-and-pony show for more than a hundred reporters, from Japanese TV crews to British tabloid reporters to 60 *Minutes.* *

I had read countless articles about Doron as a tyke, all filled with his quick wit and his insufferable boasting. I had also seen how his mother had made a spectacle of him, sharing intimate facts about his life with millions of readers and viewers. I wanted to know what Doron had become as an adult and what it was like to be the Repository's symbol. In spring 2001, I tracked him down at Reed College, where he was an eighteen-year-old finishing his freshman year. After I had left several messages on his sitar-twanging answering machine, we finally talked.

"I was [Robert Graham's] emblem. I was the boy with the high IQ who was not screwed up. I was his ideal result."

I had seen pictures of Doron: he was a goateed, gentle-looking hippie. His voice, however, was suffused with ennui and bitterness.

---

* Charging for interviews is not an unreasonable position: Doron sees no reason to waste his life as an unpaid sideshow freak. He wants something for his inconvenience, for having to answer the same tedious questions: *So, are you a genius?* Foreign reporters give him cash without hesitation. American journalists arrange byzantine schemes to avoid paying him directly. For example, 60 *Minutes* repeatedly flew Doron's mother from Los Angeles to visit him at boarding school in New Hampshire, housing her at top-notch hotels. Some journalists have "rented" Afton's house in order to obtain an interview or booked her for one of her psychotherapy sessions. Other media outlets don't disclose that they've done this.

I managed to get Afton to talk to me by taking her out to a nice dinner in Pasadena. Doron was more complicated. When I reached him in 2001, he asked for cash. I refused, but, at my *Slate* editor's suggestion, I proposed that the magazine fly him to the East Coast so that he could visit his best friend in New Hampshire and I could meet him. That way he would get a benefit he wanted, a plane ticket worth several hundred dollars, and I could feel as if I were not buying him off. My convoluted ethical justification: I would have had to fly cross-country to see him, so what difference did it make if the airplane fare went to me or him?

He agreed, but it turned out he couldn't fly east until months after I needed to talk to him. So I interviewed him over the phone for a long time and made plans to meet him at the airport for a follow-up when he finally came east. But because his schedule changed, I never did meet him face-to-face.

The reason for the ennui was obvious: he had delivered this spiel many times before, and he was sick of it. (Doron, who usually stuttered, didn't when he was talking about sperm. Maybe he was too well rehearsed.) But the bitterness came from somewhere else. He said "ideal result" with derision. When Doron was a boy and his mother, Afton, was thrusting him in front of the cameras, he was the hero of the Nobel sperm bank. Now that he was an adult and controlled his story, he was giving it a different ending.

"It was a screwed-up idea, making genius people," he said. "The fact that I have a huge IQ does not make me a person who is good or happy. People come expecting me to have all these achievements under my belt, and I don't. I have not done anything that special.

"I don't think being intelligent is what makes a person. What makes a person is being raised in a loving family with loving parents who don't pressure them. If I was born with an IQ of 100 and not 180, I could do just as much with my life. I don't think you can breed for good people."

Both Afton Blake and Doron insist that she never pressured him into his youthful achievements. She was an indulgent mother, but she wasn't a stage mom. Doron discovered by himself that he was a math prodigy and a wonderful musician. He shone at a Los Angeles school for the gifted, then won a scholarship to Phillips Exeter Academy in New Hampshire, one of the best high schools in the nation.

But if Afton didn't coerce Doron into achieving, she did something worse. She turned her son's life into *The Truman Show*. British tabloid journalists visited his dorm. His love life was bandied about in print; also his difficulty making friends. His accomplishments were national news. Doron loved his mom enormously, but he had come to realize how his public childhood had twisted his life. Doron's story was supposed to have been about nature, about his Nobel-sperm-bank-derived genetic gifts. But as Doron told it, he made it clear that he thought it was about nurture. "It was not the best thing for me to grow up in the spotlight. This is something I realized recently. I never enjoyed the media appearances, and I did not really understand the effects on me till now," he said.

"I have always been a shy spend-time-alone kind of person. Being in the public has made me very uncomfortable. It is one reason why now I feel that people are not going to like me. I always feel like people are examining me and probing me. It is much better for kids to grow up in a safe environment. It would have been much better if Mom had not had me microprobed.

"Most of being a prodigy was negative," he continued. "People have always been saying 'prodigy sperm child' all my life. But I am not that wonderful at anything. You feel a lot of pressure because you don't want to let people down, or you don't really feel free to be what you want to be.

"Mom did not mean to, but she put a burden on me by making me feel like someone special," he once said. "I'm always hearing that I'm special. I don't want to be special."

Doron told me he believed he was "smart" in the sense that he processed information quickly. He did think that was genetic. He also thought it didn't matter. In fact, he seemed to be going out of his way to avoid using that intelligence. Since he'd started college, he had abandoned math and science, the subjects he excelled at. He was intending to major in comparative religion. He was also passionate about music—fluent on piano, guitar, and sitar—but apprehensive about playing in public. The only career he could imagine for himself was teaching at his old high school, Phillips Exeter— "where brilliant kids have brilliant thoughts." Maybe this was just the loneliness of freshman year speaking. Still, his hope of a return to Exeter seemed poignant. He wanted to go back in the place where he had been safest and happiest.

Doron didn't exactly resent his sperm bank birth. One of the first things he said to me, in fact, was that the reason he did interviews was that he wanted to show people that sperm bank kids were just like everyone else. Still, he was remarkably uncurious about his donor. He said the BBC had approached him a couple of years earlier and told him it had figured out the identity of Donor Red #28, his father. They asked Doron if he wanted to meet him. Doron said he told them that he would meet him but didn't really care. How little did this matter to him? Doron claimed, and I believe him, that he had forgotten Red 28's name after the BBC had told him. "I think it was John, and he was a computer scientist of some sort.

"Genes have never been important to me. Family is the people you love. I feel a lot closer to people who are not my blood than to those who are. Those blood ties have never been enough to hold me ever. [The donor] is not part of my life. He has no place in my life whatsoever. He is no more than a stranger."

It is hard to imagine what Robert Graham would have made of his favorite offspring, now that he was all grown up. Graham prized science and scorned emotion. He hoped his sperm bank kids would build computers and synthesize medicines. Doron, once the math whiz, had disavowed hard science for the softest of studies, human spirituality. Graham had dismissed his own youthful musical career as a "waste"; Doron lived through his music. Graham sought athletic, macho donors. Doron despised sports and couldn't stand manly men. Graham hoped his sperm bank kids would lead the world. Doron's ambition was to teach high school. Doron was everything that Graham dreamed of—hyperarticulate, smart, brutally honest—yet he rejected all that Graham preached about genetics and intelligence. The power of Doron's brain vindicated Graham. The feeling in Doron's heart rejected him.

EPILOGUE
SEPTEMBER 2004

When Tom returned home from visiting Jeremy in Florida, he expected that he and his new dad would stay close. Tom was gratified when Jeremy immediately kept one promise he had made to Tom in Miami: Jeremy tracked down the forms that Lana needed for her green card application and FedExed them to Kansas City. Jeremy also called Lana on her birthday in November and sent Tom $100 for Christmas. But in other ways, in the ways that mattered to Tom, he and Jeremy drifted apart. They didn't talk much, a phone call or two per month. When they did, the conversations were comfortable but shallow. Jeremy reneged on his promise to visit Kansas City in the fall. He said he would come around New Year's, but that didn't happen, either. Instead, Jeremy pushed the visit back a few more months. Maybe he'd come in May, he said. By late January, Jeremy had stopped calling Tom. He was still friendly when Tom called him, but Tom got the message: Jeremy's interest was sagging.

Tom had come to recognize that Jeremy, who had neglected most of his own kids, wasn't capable of being his dad, either. "Before

I left Florida, I said to myself, 'Tom, be honest with yourself, he is not going to be a father.' And he shouldn't be. That is not what he signed up to do. I am hoping we can have a close friendship, but that is the most it will ever be. And now, if I have a problem, I can't see going to him to talk about it. He doesn't feel like family, at least not yet. Really, I am lucky even to have met him, so I shouldn't expect anything like that."

Since meeting Jeremy, Tom had stopped talking about a subject that had used to fascinate him: genetic potential. In the beginning, Tom had been mesmerized by his genetic heritage, and destiny: *Jonas Salk, my dad?* After Mary had told him he was a Nobel sperm baby, Tom had started to believe in DNA, to hope that his fate was in his genes. But now he had discovered that his genetic benefactor wasn't a Nobelist, wasn't a genius, wasn't even an admirable man. Having accepted the DNA religion, he was ready to abandon it. When he had started searching for his Nobel sperm donor dad, Tom had hoped that genes were destiny. Since he had found Jeremy, he had hoped they were not.

Through 2004, Jeremy slipped further and further into the background of Tom's manic life. Baby, wife, mom, job, school: Tom had to deal with all of them; all of them were exhausting. During the winter, Tom's relationship with his mom fell apart. They had been sniping at each other for months, partly because Tom disliked her new boyfriend. When Mary and the boyfriend got engaged, Tom and his mom stopped talking. Tom and Lana were still living in Mary's basement, but he started making plans to move out as soon as he could save enough for a security deposit.

Meanwhile, the rest of Tom's life settled into a kind of order. He was impossibly busy, but he was not unhappy. He was on the verge of completing his two-year degree and was planning what to study for his B.A. — maybe computer science. Lana was going to start classes in massage therapy, and she was also moving steadily toward her green card. In the spring, Tom and Lana decided to celebrate the wedding they had never had when they'd gotten married—a joyous, drunken Russian blowout at Lana's parents' house. They planned it for May, and Tom invited Jeremy. Jeremy said he would come, but of course he didn't. Tom enjoyed himself so much at the party that he didn't really care that his "real dad" was missing.

Oddly, Tom found himself getting closer and closer to his old dad, Alvin. Seldom happy at the best of times, Alvin had been in a funk since his divorce from Mary. Sadness had replaced his usual simmering anger. He sought Tom's comfort, and Tom was glad to give it; it was the first warm emotional exchange they had had in years. When his dad came home from the road, he slept on Tom and Lana's couch. He even paid a little attention to Darian.

It puzzled Tom that his relationship with his mother and Jeremy—his biological parents—would fail while he loved his dad and Lana more and more. He e-mailed me recently, "I know my dad isn't really my dad, yet I'm a hundred times closer with him than I am with Jeremy. I know my mom really is my mom and I have a blood relation to her, and yet I am much closer to my dad right now despite the lack of biological relation. And I am closer to Lana than I am with anyone else and I am not related to her in any way shape or form. In the end I don't think that your family matters that much in the sense of who they are. I think it matters who—in and out of your family—you embrace and accept as your family."

Tom reached out for one more embrace. He had not talked to his half brother Alton Grant for more than a year. But in late summer of 2004, Tom e-mailed Alton again. Alton quickly replied, and they started trading messages. They talked about music, their families, school. It made Tom happy to connect with his brother again. Alton mentioned to Tom that he had refused to meet Jeremy when Jeremy had shown up at his house. Tom, who is usually shy of giving advice, told me he couldn't help it this time. He told Alton about how his own relationship with Jeremy had soured, but then he suggested that Alton see Jeremy anyway. Just in case, Tom urged, you should meet your father. Just in case.

One night in 2003, at around eleven P.M., I got a call at home from a young woman. She said her name was Tatiana. She sounded as if she was about to cry, and then she did. I had posted my number and e-mail address on various sperm bank bulletin boards, mentioning my interest in connecting kids and donors from the Repository for Germinal Choice. I had been swamped with plaintive calls and e-mails, but only a few from people who had used the Repository. Most had used other banks and clinics. They tried me because they couldn't try anyone else. Tatiana was one of those callers.

She told me she had a two-year-old daughter through Fairfax Cryobank. "What I did was *wrong*! I should never have had this child. I was twenty-two. I just wanted it. I needed a child. The force of nature or something. I needed it. I don't know why. So I went to the cryobank, and I picked a donor. It was a mistake. As soon as I got pregnant from the sperm donor, I had nightmares. Who is that inside? Who is it? I could not bond with the baby inside. *It* was inside."

Now, Tatiana said, she had to find the donor, and that was why she was calling me. "I have been looking for him for a year. I have narrowed it to sixty names, because the bank told me the name of the medical school he went to, even the year he graduated. I have ordered the yearbook. Now you have to help me find him." Tatiana insisted to me that finding the donor would solve her problems, that it would help her bond to her daughter. I tried to dissuade her from searching. I argued that knowing the donor's name or even shaking his hand wouldn't make her daughter more her daughter. And it also wouldn't make her daughter more *his* daughter.

The more time I spent around DI parents and kids, the more I got the sense of how deep and how widespread the yearning of people like Tatiana was. They hungered both for the truth and for family. Around one million American kids are "donor offspring." Most of them don't know it, but more and more of them do, because the walls of silence around AID are collapsing. Single women and lesbians are pushing for openness, and so are married couples who have used DI. Eventually, sperm donor anonymity will end in the United States, as it is ending in Europe. Anonymity will end for the same reason adoption records are being unsealed. The laws that protected adoption and sperm bank records were drafted in a pre-DNA era. But the past thirty years have witnessed an explosion of knowledge about DNA, from the Human Genome Project to gene therapy to preimplantation genetic diagnosis. Increasingly we accept that our genes guide us and even rule us. That triumph of geneism has enormous implications for AID and adoption: once you decide that you are defined by your genes, it implies a right to know your genetic history.

But I'm not sure we are ready for the end of anonymity yet. In my journeys with the families, children, and donors of the Nobel sperm bank, I came to believe that even if there's a moral (and perhaps one day, legal) right to know a sperm donor father, that won't necessarily create happiness or psychological closure—or a family. A sperm donor, after all, is a particularly *unpromising* candidate for a relationship with his child. Adopted children know that at least *something*—an event as transient as a one-night stand or as durable as a marriage—happened between their birth parents. And they know that their mother carried them for nine months, gave birth to them, and held them for at least a moment, and probably much more. But a sperm donor offers no such consolation. There is no bond, no physical connection between father and child. Tom once told me he wanted to know what Jeremy had been thinking when he donated. What *was* Jeremy thinking when he "conceived" Tom? Probably about the breasts on the model in the *Playboy* he was thumbing. Today, there is so much expectation about these sperm donors, so much hope that they will turn out to be donor "dads." Yet their only connection to their "children" is that once upon a time, maybe on a grubby couch, they masturbated into a sterile plastic cup. Fatherhood is not what they signed up for; it's not what they want; it's probably not anything they would be good at. Sometimes, faced with donors like Jeremy or Michael the Nobelist's son, I wondered if it would be easier if we regarded donors as mere equipment, rather than dreamed about them as fathers.

But then I thought about Donor White. It was more than two years after Roger had first written to me. The gloom that had settled on his life during his illness had totally lifted. He was e-mailing with Joy a couple of times a week and still found her "highly pleasing." Beth always sent Roger a copy of Joy's report card, knowing that no one would be prouder of it than he. For Father's Day 2003, the anniversary of Beth's initial e-mail to Roger, Joy gave Roger the needlepoint sampler she had been working on when I had seen her months earlier. The sampler was a poem, the first letter of each line decorated with red roses:

The kiss of the sun for pardon
The song of the birds for mirth
One is nearer God's heart in a garden
Than anywhere else on earth

Joy Lily, June 15, 2003

A card accompanied the sampler: "I hope you like your present because every stitch was made with happy thoughts of you. Lots of love, Joy."

"Needless to say," Roger told me, "it is one of my most prized possessions."

Father's Day 2004 brought an even better gift: Joy and Beth visited Roger again, and Joy's stepdad even came along. And Roger made his own trip to her, his first travel since his illness. He and Rebecca flew to Pennsylvania to see Joy and tour Civil War battlefields. Roger also dreamed of taking Joy with him on a visit to Virginia, so they could see where their colonial ancestors had settled more than three hundred years before.

Roger had undertaken an ambitious renovation project, too. His house had been falling apart for years, and once he had gotten sick, he had assumed he would never do anything about it: Why bother? But Joy had been talking about studying marine biology. The Scripps Oceanography Institute was just a few miles from Roger and Rebecca's house; they got it into their heads that Joy might enroll there someday. So they added a second floor to the house, including a study for Joy. They included an extra bedroom, too: maybe she could live with them.

Not long ago, Roger wrote me that the more he came to know Joy, the less he thought about his scientific achievements—the piles of articles and patents that had consumed his life. "When I first had my health problems and knew nothing of Joy, I felt like I would be leaving nothing behind and that my life had been devoted largely to work that had amounted to very little. Now I feel that there will be something worthwhile left behind, and that thought is a comfort to me. . . . She will serve as my continuation into the future after my days on this earth are done."

I noticed something else, too. The more Roger came to know Joy, the less he cared about his own intelligence and the less he cared about hers. He rarely mentioned the Repository to me anymore. He hadn't forgotten it, but he had set it aside. Dr. Graham's grand, strange experiment, the fact that Roger belonged to an elite cadre of donors, the fact that Joy had been designed for greatness—these no longer interested him. There was no more experiment, only a child. Whether she was a genius, whether she could change the world—who cared? She was happy, and she was his. That was all that mattered.

ACKNOWLEDGMENTS

The children, parents, and donors of the Repository for Germinal Choice took a risk by sharing their secrets with me. I can't thank them enough. *The Genius Factory* wouldn't exist without them. I wish I could name them, but pseudonyms will have to do. Thank you Tom and Mary Legare, Samantha and Alton Grant, Jeremy Sampson, Donor White, Beth, and Joy. I was amazed at the grace with which you handled events that were sometimes joyful, often awkward, and always complicated.

I owe my colleagues at *Slate* big time. Michael Kinsley pushed me to investigate the Nobel sperm bank, and Jacob Weisberg, his successor as *Slate*'s editor, encouraged me to persist with the project. Jack Shafer came up with the inspired idea of using the Internet to let children and donors find me. June Thomas thought up a great name for my series, "Seed." Practically everyone else at *Slate* contributed valuable leads, suggestions, and corrections for the project.

My agent, Rafe Sagalyn, had the sense to recognize that the Nobel sperm bank was a book, even when I didn't. Jonathan Karp at Random House has the gift of the true editor: he knows that you have to edit something *before* it's written. When I was floundering around with fifteen different ideas for the book, Jon identified the real story and gave me a great big shove in the right direction. His colleagues at Random House, particularly Jonathan Jao and Jillian Quint, were helpful at every stage, and Lynn Anderson and Dennis Ambrose saved me from dreadful mistakes.

Derek Anderson was my sounding board, my private eye, and my pal. Derek, who was researching a documentary about the Repository as I was working on my book, was unbelievably generous with his time and his discoveries. I never would have unraveled the story of Donor Coral without him.

I don't have the space to thank everyone who sat for interviews or contributed to my research, but I'm particularly grateful to the following people:

Doron and Afton Blake, the Ramm family, Lorraine O'Brien, Edward Burnham, Donor Light Blue, Donor Orange, Donor Blue/Black, Donor Aqua, and many other donors and families gave me great insight into why mothers used the Repository, why donors contributed to it, and how kids felt about it.

Marta Ve Graham supplied me with a useful documentary record of her husband Robert Graham's life. Robert Graham's son Robin, brother Tom, nephews Tom and Jeff, and daughter-in-law Diane all shared memories of Robert Graham with me.

Former Repository employees Paul Smith, Julianna McKillop, and Anita Neff spent hours describing the inner workings of the sperm bank. Steve Broder filled me in on the early history of the Repository, and was a trove of information about modern sperm banking. Dr. Cappy Rothman and Marla Eby told me all about California Cryobank, the gold standard of American sperm banks. Joanna Scheib tutored me on the identity release program at the Sperm Bank of California. Wendy Kramer explained how her Donor Sibling Registry works. Lori Andrews and Alexander Capron advised me on the ethics of the fertility industry. Margaret Williams guided me through the marvelous William Shockley archive at

Stanford. Edwin Chen of the *Los Angeles Times* recalled the media circus that followed his 1980 story about the Repository. At the Genetics and Public Policy Center, Susannah Baruch, Gail Javitt, and Kathy Hudson instructed me on the law and science of fertility. Connie Cappel brought the history of Harbor Springs to life. She also put me in touch with Bruce Gathman, who debunked the myth of Ephraim Shay. Shannon Brownlee sent me some great articles on imprinting. Derek Anderson's colleagues at Cinenova, particularly Ric Bienstock, Kathryn Liptrott, and David Lint, helped by sharing material they had gathered for their documentary. *Saturday Night Live* historian Patrick Lonergan tracked down the "Dr. Shockley's House of Sperm" sketch. I consulted lots of books: The three most useful (and fascinating) were Daniel Kevles's *In the Name of Eugenics*, Gina Maranto's *Quest for Perfection*, and Lillian Hoddeson and Michael Riordan's *Crystal Fire*.

Sarah Stillman, Stephen Baxter, and Rebecca Gordon were superb research assistants. I wish I could have paid you more!

All of my friends tolerated my endless sperm bank stories graciously. Some did even more. Marjorie Williams suggested the structure of the book. Frank Foer, Linda Perlstein, Emily Yoffe, and Ben Wittes gave me excellent advice about how to write a book. Arthur Allen and Christine Rosen told me some of their favorite eugenics stories. Michael Raunitzky did some nifty online detective work for me. David Greenberg proposed a subtitle. Ben Sheffner, David and Nancy Sheffner, Tonje Vetleseter, Auran Piatagorski, Michael and Kathryn Koehler, Craig and Anne Turk, Ben Cooley, and Jenny Konner offered warm beds and hot meals during my various California sojourns. Rosemary Quigley counseled me on the ethics of sperm banking, and pointed me at some great legal cases. Rosemary died the week I finished the book: I wish she had lived to argue with me about it.

Clara Jeffery, Margaret Talbot, and David Finkel did the kindest, most masochistic thing you can do for a friend: They read the first draft. Their comments were incredibly helpful in clarifying and shaping the book.

My brother John Plotz offered tons of encouragement, as did my in-laws Miriam and Eli Rosin, Michael Rosin, Dalila Rosin, and Lisa Soltani.

This is a book about what it means to be a parent. Fortunately, I learned about that from the best: My parents, Paul and Judith Plotz, have been all that a son could hope for—loving, attentive, curious, hands-off, hands-on, funny. I never would have known about the Nobel sperm bank if not for my father. I never would have had the wit to write about it if not for my mother.

My daughter, Noa, was born just as I started researching the Nobel sperm bank, and my son, Jacob, was born just as I started writing the book. In *The Genius Factory*, I have tried to explain what it feels like to be a father: I wouldn't have understood anything about that without Noa and Jacob.

This book is for my wife, Hanna. She made me write it when I didn't have the guts to do it. She's my love.

PHOTO: © HANNA ROSIN

DAVID PLOTZ is deputy editor of the online magazine *Slate*. He lives in Washington, D.C., with his wife and two children. For more information on *The Genius Factory*, including original documents, please visit www.thegeniusfactory.net.